AF518279

Biological Control of Natural Enemies

The Author

Dr. Admasu Gelana, Senior Expert in Crop Protection at Adet Agricultural Research Center, Bahidar, Ethiopia. Dr. Gelana has published more than 85 research papers in International journals of repute. Besides this he has participated/delivered lectures at various International Conferences/Symposia etc. Presently, he is working as Consultant (Crop Protection) at Bahidar, Ethiopia.

Biological Control of Natural Enemies

Dr. Admasu Gelana

2026

Agri Horti Press

New Delhi-110 084

© 2026 Reserved

ISBN: 978-93-83285-54-9 (Hardbound)

Publisher's note:

All Rights reserved under International Copyright Conventions. No part of this publication may be reproduced, stored in a retrieval system, or transmitted in any form or by any means, electronic, mechanical, photocopying, recording or otherwise without the prior written consent of the publisher and the copyright owner.

Published By

Agri Horti Press

New Delhi-110 084

Email: agrihortipress@gmail.com

Website: www.agrihorti.com

Digitally Printed at : Replika Press Pvt. Ltd.

Preface

With the advent of synthetic insecticides in the 1950s, easy control of insect pests appeared at hand. However, it soon became obvious that there were problems associated with the use of insecticides. Some insect pests became resistant, and some nontarget organisms were adversely affected, and pest resurgence occurred. Additionally, environmental and health concerns arose.

Today, the protection of food and fiber crops from insect, mite, disease, and weed pests in conventional agricultural systems still relies primarily on the use of chemical pesticides. However, continued reliance solely on conventional pesticides has drawbacks. The integrated pest management strategy described in this book promotes non-chemical pest control tactics such as use of pest-resistant plants, cultural control methods, and biological control. Pesticides should be used only to prevent an economic loss and rarely should be used in a prophylactic manner.

The need to develop alternatives to conventional pesticides may be more acute in some commodities than in others. For example, vegetables and fruits are considered minor crops in many areas, and new insecticides are less likely to be registered or existing ones re-registered. Populations of many major vegetable and fruit insect pests possess resistance to insecticides. Also, vegetable and fruit growers, especially fresh market producers with small and diverse operations and roadside stands, are highly visible to the public. The application of pesticides is frequently obvious and may result in conflicts with urban neighbors. Alternatives that would reduce the need for pesticide use could help alleviate some of these conflicts.

All insect pests have natural enemies. The use of these organisms to manage pests is known as biological control. Conservation of natural enemies is probably the most important biological control practice readily available. Through the use of selective insecticides and the judicious use of broader spectrum materials, natural enemies can exist and exert their impact on pest populations. The future of biological control is promising, but this tactic will constitute just one of many pest management options. Many obstacles will need to be overcome before biological control can reach its full potential.

Traditionally, educational and extension efforts have not provided growers with the necessary information to help identify natural enemies or determine their presence in fields. The importance of natural enemies cannot be over-emphasized, and making educators (e.g., extension agents) and growers aware of their existence would be a first and very important step — a primary purpose of this guide.

Before biological control will advance, much more emphasis needs to be placed on investigating indigenous natural enemies and their impact on the pests they attack. With this information it may be possible to foster or enhance the efficacy of natural enemies through manipulation of the crop habitat, changes in cultural practices, or changes in pesticide application practices. In addition, the introduction of new natural enemies through classical biological control programs holds much promise. The development of successful biological control programs will be challenging, but holds great potential.

This book provides descriptions of biological control (or biocontrol) agents of insect, disease, and weed pests. It is also a tutorial on the concept and practice of biological control and integrated pest management (IPM). Whether you are an educator, a student, a researcher, or an extension or regulatory agent, I hope you will find this information useful. The book currently includes approximately 100 natural enemies of pest species. On each of these you will see descriptions of the life cycles and habits, and other useful information about each natural enemy. Four types of natural enemies are included in this book, i.e. parasitoids, pathogens, predators and weed-feeders.

I hope the book serves a good source of information to students, teachers, researchers, extension personnel, educators, land managers enaged in the field of agriculture.

Though, I have taken great care, I would be grateful to the readers who may point out any errors and useful suggestions.

Dr. Admasu Gelana

Contents

1

Introduction

This segment includes several paragraphs with general information about biological control and these subsections:

- Conservation
- Classical Biological Control
- Augmentation
- Purchase and Release of Natural Enemies

Biological control is a component of an integrated pest management strategy. It is defined as the reduction of pest populations by natural enemies and typically involves an active human role. Keep in mind that all insect species are also suppressed by naturally occurring organisms and environmental factors, with no human input. This is frequently referred to as natural control. This guide emphasizes the biological control of insects but biological control of weeds and plant diseases is also included. Natural enemies of insect pests, also known as biological control agents, include predators, parasitoids, and pathogens. Biological control of weeds includes insects and pathogens. Biological control agents of plant diseases are most often referred to as antagonists.

Predators, such as lady beetles and lacewings, are mainly free-living species that consume a large number of prey during their lifetime. Parasitoids are species whose immature stage develops on or within a single insect host, ultimately killing the host. Many species of wasps and some flies are parasitoids. Pathogens are disease-causing organisms including bacteria, fungi, and viruses. They kill or debilitate their host and are relatively specific to certain insect groups. Each of these natural enemy groups is discussed in much greater detail in following sections.

The behaviors and life cycles of natural enemies can be relatively simple or extraordinarily complex, and not all natural enemies of insects are beneficial to crop production. For example, hyperparasitoids are parasitoids of other parasitoids. In potatoes grown in Maine, 22 parasitoids of aphids were identified, yet these were attacked by 18 additional species of hyperparasitoids.

This guide concentrates on those species for which the benefits of their presence outweigh any disadvantages. A successful natural enemy should have a high reproductive rate, good searching ability, host specificity, be adaptable to different environmental conditions, and be synchronized with its host (pest).

A high reproductive rate is important so that populations of the natural enemy can rapidly increase when hosts are available. The natural enemy must be effective at searching for its host and it should be searching for only one or a few host species. Spiders, for example, feed on many different hosts including other natural enemies. It is also very important that the natural enemy occur at the same time as its host. For example, if the natural enemy is an egg parasitoid, it must be present when host eggs are available. No natural enemy has all these attributes, but those with several characteristics will be more important in helping maintain pest populations.

There are three broad and somewhat overlapping types of biological control: conservation, classical biological control (introduction of natural enemies to a new locale), and augmentation.

THE NEED, STATUS, AND POTENTIAL FOR BIOLOGICAL CONTROL

With the advent of synthetic insecticides in the 1950s, easy control of insect pests appeared at hand. However, it soon became obvious that there were problems associated with the use of insecticides. Some insect pests became resistant, and some nontarget organisms were adversely affected, and pest resurgence occurred. Additionally, environmental and health concerns arose.

Today, the protection of food and fiber crops from insect, mite, disease, and weed pests in conventional agricultural systems still relies primarily on the use of chemical pesticides. However, continued reliance solely on conventional pesticides has drawbacks. The integrated pest management strategy described in this guide promotes nonchemical pest control tactics such as use of pest-resistant plants, cultural control methods, and biological control. Pesticides should be used only to prevent an economic loss and rarely should be used in a prophylactic manner.

The need to develop alternatives to conventional pesticides may be more acute in some commodities than in others. For example, vegetables and fruits are considered minor crops in many areas, and new insecticides are less likely to be registered or existing ones re-registered. Populations of many major vegetable and fruit insect pests possess resistance to insecticides. Also, vegetable and fruit growers, especially fresh market producers with small and diverse operations and roadside stands, are highly visible to the public. The application of pesticides is frequently obvious and may result in conflicts with urban neighbors. Alternatives that would reduce the need for pesticide use could help alleviate some of these conflicts.

All insect pests have natural enemies. The use of these organisms to manage pests is known as biological control. Conservation of natural enemies is probably the most important biological control practice readily available. Through the use of selective insecticides and the judicious use of broader spectrum materials, natural enemies can exist and exert their impact on pest populations. The future of biological control is promising, but this tactic will constitute just one of many pest management options. Many obstacles will need to be overcome before biological control can reach its full potential.

Traditionally, educational and extension efforts have not provided growers with the necessary information to help identify natural enemies or determine their presence in fields. The importance of natural enemies cannot be over-emphasized, and making educators (e.g., extension agents) and growers aware of their existence would be a first and very important step — a primary purpose of this guide.

Before biological control will advance, much more emphasis needs to be placed on investigating indigenous natural enemies and their impact on the pests they attack. With this information it may be possible to foster or enhance the efficacy of natural enemies through manipulation of the crop habitat, changes in cultural practices, or changes in pesticide application practices. In addition, the introduction of new natural enemies through classical biological control programs holds much promise. The development of successful biological control programs will be challenging, but holds great potential.

CONSERVATION

The conservation of natural enemies is probably the most important and readily available biological control practice available to growers. Natural enemies occur in all production systems, from the backyard garden to the commercial field. They are adapted to the local environment and to the target pest, and their conservation is generally simple and cost-effective. With relatively little effort the activity of these natural enemies can be observed.

Lacewings, lady beetles, hover fly larvae, and parasitized aphid mummies are almost always present in aphid colonies. Fungus-infected adult flies are often common following periods of high humidity. These natural controls are important and need to be conserved and considered when making pest management decisions. In many instances the importance of natural enemies has not been adequately studied or does not become apparent until insecticide use is stopped or reduced. Often the best we can do is to recognize that these factors are present and minimize negative impacts on them. If an insecticide is needed, every effort should be made to use a selective material in a selective manner.

CLASSICAL BIOLOGICAL CONTROL

In many instances the complex of natural enemies associated with an insect pest may be inadequate. This is especially evident when an insect pest is accidentally introduced into a new geographic area without its associated natural enemies. These introduced pests are referred to as exotics and comprise about 40% of the insect pests in the United States. Examples of introduced vegetable pests include the European corn borer, one of the most destructive insects in North America. To obtain the needed natural enemies, we turn to classical biological control. This is the practice of importing, and releasing for establishment, natural enemies to control an introduced (exotic) pest, although it is also practiced against native insect pests. The first step in the process is to determine the origin of the introduced pest and then collect appropriate natural enemies (from that location or similar locations) associated with the pest or closely related species. The natural enemy is then passed through a rigorous quarantine process, to ensure that no unwanted organisms (such as hyperparasitoids) are introduced, then reared, ideally in large numbers, and released. Follow-up studies are conducted to determine if the natural enemy successfully established at the site of release, and to assess the long-term benefit of its presence.

There are many examples of successful classical biological control programs. One of the earliest successes was with the cottony cushion scale, a pest that was devastating the California citrus industry in the late 1800s. A predatory insect, the vedalia beetle, and a parasitoid fly were introduced from Australia. Within a few years the cottony cushion scale was completely controlled by these introduced natural enemies. Damage from the alfalfa weevil, a serious introduced pest of forage, was substantially reduced by the introduction of several natural enemies. About 20 years after their introduction, the alfalfa acreage treated for alfalfa weevil in the northeastern United States was reduced by 75 percent. A small wasp, Trichogramma ostriniae, introduced from China to help control the European corn borer, is a recent example of a long history of classical biological control efforts for this major pest. Many classical biological control programs for insect pests and weeds are under way across the United States and Canada.

Classical biological control is long lasting and inexpensive. Other than the initial costs of collection, importation, and rearing, little expense is incurred. When a natural enemy is successfully established it rarely requires additional input and it continues to kill the pest with no direct help from humans and at no cost. Unfortunately, classical biological control does not always work. It is usually most effective against exotic pests and less so against native insect pests. The reasons for failure are often not known, but may include the release of too few individuals, poor adaptation of the natural enemy to environmental conditions at the release location, and lack of synchrony between the life cycle of the natural enemy and host pest.

AUGMENTATION

This third type of biological control involves the supplemental release of natural enemies. Relatively few natural enemies may be released at a critical time of the season (inoculative release) or literally millions may be released (inundative release). Additionally, the cropping system may be modified to favor or augment the natural enemies. This latter practice is frequently referred to as habitat manipulation.

An example of inoculative release occurs in greenhouse production of several crops. Periodic releases of the parasitoid, Encarsia formosa, are used to control greenhouse whitefly, and the predaceous mite, Phytoseiulus persimilis, is used for control of the two-spotted spider mite.

Lady beetles, lacewings, or parasitoids such as Trichogramma are frequently released in large numbers (inundative release). Recommended release rates for Trichogramma in vegetable or field crops range from 5,000 to 200,000 per acre per week depending on level of pest infestation. Similarly, entomopathogenic nematodes are released at rates of millions and even billions per acre for control of certain soil-dwelling insect pests.

Habitat or environmental manipulation is another form of augmentation. This tactic involves altering the cropping system to augment or enhance the effectiveness of a natural enemy. Many adult parasitoids and predators benefit from sources of nectar and the protection provided by refuges such as hedgerows, cover crops, and weedy borders.

Mixed plantings and the provision of flowering borders can increase the diversity of habitats and provide shelter and alternative food sources. They are easily incorporated into home gardens and even small-scale commercial plantings, but are more difficult to accommodate in large-scale crop production. There may also be some conflict with pest control for the large producer because of the difficulty of targeting the pest species and the use of refuges by the pest insects as well as natural enemies.

Examples of habitat manipulation include growing flowering plants (pollen and nectar sources) near crops to attract and maintain populations of natural enemies. For example, hover fly adults can be attracted to umbelliferous plants in bloom.

Recent work has demonstrated that planting prune trees in grape vineyards provides an improved overwintering habitat or refuge for a key grape pest parasitoid. The prune trees harbor an alternate host for the parasitoid, which could previously overwinter only at great distances from most vineyards. Caution should be used with this tactic because some plants attractive to natural enemies may also be hosts for certain plant diseases, especially plant viruses that could be vectored by insect pests to the crop. Although the tactic appears to hold much promise, only a few examples have been adequately researched and developed.

PURCHASE AND RELEASE OF NATURAL ENEMIES

Many commercial insectaries rear and market a variety of natural enemies including predaceous mites, lady beetles, lacewings, praying mantids, and several species of parasitoids. Success with such releases requires appropriate timing (the host must be present or the natural enemy will simply die or leave the area) and release of the correct number of natural enemies per unit area (release rate). In many cases, the most effective release rate has not been identified as it will vary depending on crop type and target host density.

Success also requires a healthy and robust natural enemy. This book does not make specific recommendations about the purchase or release of the commercially available natural enemies, but it does provide essential information about the biology and behavior of most commercially reared species. This information should be helpful in making decisions regarding their use.

Conservation of natural enemies is arguably the most important concept in the practice of biological control and fortunately is also one of the easiest to understand. Simply put, conservation of natural enemies means avoiding practices which are harmful to natural enemies and implementing practices which benefit them. Sounds like good common sense! The tricky part comes in understanding exactly what practices are harmful and how beneficial practices can be integrated into your production system. That requires that we understand the biology of the natural enemies and are willing to modify our practices to accommodate them.

Natural Enemies as Livestock

Everyone understands that a dairy cow needs food, water, shelter and protection from adverse conditions. To perform her best she also needs protection from biting flies, diseases and in some cases, predators which may injure or kill her. The dairy producer knows her requirements change through the year and makes provisions to provide for these needs. In the winter, shelter is critical, while in the summer adequate water and shade are necessary. In some months, grazing may provide her total food requirements, but as pasture growth slows, supplemental food may need to be provided.

Natural enemies have exactly the same types of needs as the dairy cow. To perform their best, they need food, shelter and protection from adverse conditions as well. Frequently, we do not fully understand or make allowances to provide for these needs. The result is many instances where biological control could be very effective, but has failed or resulted in less than adequate control because we did not provide for the natural enemies' basic requirements. So what do natural enemies need and how can we help them?

Avoid Harmful Practices

The most obvious practice to avoid is the use of insecticides at times when natural enemies will be harmed. Insecticides can have direct effects on natural enemies by poisoning them or indirect effects by eliminating their hosts and causing starvation. In some cases, insecticides can be successfully integrated into the system without causing harm to natural enemies. This may be through the use of a selective insecticide such as Bt, timing the application to avoid times of the day or season when important natural enemies would be exposed, or placement of the insecticide in a location where natural enemies will not contact it. In other cases, adequately protecting natural enemies may require the elimination of insecticide use.

Certain cultural practices can also be detrimental to natural enemies. Plowing, cultivation, mowing or harvesting operations which disrupt natural enemies at critical points in their life cycle should be avoided. Excessive amounts of dust from roads or cultural operations can also disrupt the activities of predators and parasitoids resulting in reduced control. Burning of crop residues or inappropriately timed irrigation can also kill many natural enemies. Finally the ambiguous category of "clean farming" which includes removal of weeds and noncrop habitats has been identified as detrimental to many natural enemies.

Incorporate Beneficial Practices

Here is where a detailed understanding of the biology of the important natural enemies in your system becomes extremely critical. If you do not know what natural enemies you want to manage, it is doubtful that you will be successful. So the first step is to gather information on the types of natural enemies you want to conserve. Then consider these points:

- Where does the natural enemy overwinter? In England, a group of researchers discovered that important predators of aphids in wheat overwintered in areas of tussock grasses in hedgerows on the edges of fields. The predators migrated into the fields in the spring, but got there too late to control aphids it the center of the fields. By planting a one meter strip of tussock grasses in the center of the field, overwintering predator numbers soared and aphid damage was controlled.

What alternate food sources do my natural enemies need? Are these present at the right times and close to my field? After emerging from overwintering, spotted lady beetles feed on plant pollen (dandelion, spring beauty, etc.) for several weeks before moving into alfalfa and wheat fields to feed on aphids. Many parasitoids also require the protein-rich pollen in order to develop new eggs. Sources of sugar (carbohydrate) are needed by many parasitoids, which they frequently obtain from the nectar of flowering plants or from aphid honeydew. Having a diversity of plants in and around fields has been shown in many cases to improve biological control.

Do my natural enemies need alternative prey/hosts? Many predators and parasitoids require alternative hosts at some points in their life cycle. *Lydella thompsoni* is a tachnid fly which parasitizes European corn borer. It emerges before ECB larvae are present in the spring and completes its first generation on common stalk borer instead. Clean farming practices which have eliminated hosts for the stalkborer are thought to have contributed to the decline of this parasitoid. Alternative prey may also be important in building up predators in a field prior to the appearance of the pest you wish them to control. Lady beetles and minute pirate bugs can consume many European corn borer eggs, but alternative prey must be present in the field prior to ECB egg laying to have maintained high predator numbers.

What shelter is needed by my natural enemies during the growing season? The activity of ground dwelling predators may be limited by high soil temperatures during the day. Incorporation of cover crops or intercrops may help to reduce soil temperatures and extend the activity period of these organisms. Similarly, many parasitoids require moderate temperatures and higher relative humidity and may need to leave fields in the heat of the day to seek shelter in shady areas. For example, the activity of predatory paper wasps increases when given wooden nest boxes which provide a protected place for them to build their nest in and near fields.

CONCLUSION

Consideration of the biological and ecological needs of natural enemies is critical for the success of any biological control effort. It is one of the easiest ways for producers to initiate biological control on their farms and should be a major consideration in any importation or augmentation program. While there are innumerable practices which may either benefit or harm the natural enemies you are seeking to manage in your production system, understanding the biology and life cycle of the specific natural enemies you want to conserve is the first step to achieving the best results.

2

The Integrated Pest Management Strategy

The management of insect pests rarely relies on a single control practice; usually a variety of tactics are integrated to maintain pests at acceptable levels. The goal of integrated pest management is not to eliminate all pests; some pests are tolerable and essential so that their natural enemies remain in the crop. Rather, the aim is to reduce pest populations to less than damaging numbers. The control tactics used in integrated pest management include pest resistant or tolerant plants, and cultural, physical, mechanical, biological, and chemical control. Applying multiple control tactics minimizes the chance that insects will adapt to any one tactic. The definition for integrated pest management most relevant to this guide comes from Flint and van den Bosch: "An ecologically based pest control strategy that relies heavily on natural mortality factors and seeks out control tactics that disrupt these factors as little as possible."

ECOLOGICAL BASIS

Integrated pest management requires an understanding of the ecology of the cropping system, including that of the pests, their natural enemies, and the surrounding environment. As discussed previously, knowledge about the ecological interrelationships between insects and their environment is critical to effective pest management. This guide emphasizes interrelationships between pest species and their natural enemies.

ECONOMIC THRESHOLD AND SAMPLING

The decision to use an insecticide, or take some other action, against an insect infestation requires an understanding of the level of damage or insect infestation a crop can tolerate without an unacceptable economic loss. The level of infestation or damage at which some action must be taken to prevent an economic loss is referred to as the "economic or action threshold." Action thresholds are available for many vegetable crops. Ideally, these thresholds adjust for changes in market prices, stage of crop growth, cost of pesticides, etc., but in reality most are based on fixed infestation or damage levels.

To estimate the severity of pest infestations, the crop or garden must be sampled. Sampling may entail examining plants and recording the number of pests or amount of damage observed, or traps may be used to capture the pest species to estimate pest abundance. Sampling is conducted at regular intervals throughout the season or during critical stages of crop growth. These sampling techniques, used either alone or in combination, estimate how close the infestation or damage level is to the economic threshold.

INTEGRATED PEST MANAGEMENT CONTROL TACTICS

- Pest-resistant crops
- Cultural Control
- Physical and Mechanical Control
- Chemical Control

Pest-Resistant Crops

One of the mainstays of integrated pest management is the use of crop varieties that are resistant or tolerant to insect pests and diseases. A resistant variety may be less preferred by the insect pest, adversely affect its normal development and survival, or the plant may tolerate the damage without an economic loss in yield or quality. Disease-resistant vegetables are widely used, whereas insect-resistant varieties are less common but nonetheless important. Examples include varieties of wheat which have tough stems that prevent development of the Hessian fly and cucurbits (squash, cucumbers, melons) that have lower concentrations of feeding stimulants (cucurbitacins) for cucumber beetles. Commercial corn lines with increased concentration of a chemical called DIMBOA display resistance to European corn borer, a major pest throughout the world. In the case of cabbage, the only reliable method of controlling onion thrips is through the use of resistant varieties.

Advantages of this tactic include ease of use, compatibility with other integrated pest management tactics, low cost, and cumulative impact on the pest (each subsequent generation of the pest is further reduced) with minimal environmental impact. The development of resistant or pest-tolerant plant varieties, however, may require considerable time and money, and resistance is not necessarily permanent. Just as insect populations have developed resistance to insecticides, populations of insects have developed that are now able to damage plant varieties that were previously resistant.

Cultural Control

There are many agricultural practices that make the environment less favorable to insect pests. Examples include cultivation of alternate hosts (e.g., weeds), crop rotation, selection of planting sites, trap crops, and adjusting the timing of planting or harvest. Crop rotation, for example, is highly recommended for management of Colorado potato beetle. The beetles overwinter in or near potato fields and they require potato or related plants for food when they emerge in the spring. With cool temperatures and no suitable food, the beetles will only crawl and be unable to fly. Planting potatoes well away from the previous year's crop prevents access to needed food and the beetles will starve. The severity and incidence of many plant diseases can also be minimized by crop rotation, and selection of the planting site may affect the severity of insect infestations.

Trap crops are planted to attract and hold pest insects where they can be managed more efficiently and prevent or reduce their movement onto valuable crops. Early planted potatoes can act as a trap crop for Colorado potato beetles emerging in the spring. Since the early potatoes are the only food source available, the beetles will congregate on these plants where they can more easily be controlled. Adjusting the timing of planting or harvest is another cultural control technique. The earlier planted processing tomatoes grown in the western United States are far less likely to be infested by the tomato fruitworm than those planted later in the season. It is also important to use pest-free transplants. Some vegetable crop transplants can be infested with insect pests, and growers using these transplants are put at a considerable disadvantage.

Physical and Mechanical Control

The use of physical barriers such as row covers or trenches prevents insects from reaching the crop. Row covers can help prevent early-season damage to cucurbits by cucumber beetles, and plastic-lined trenches are effective in trapping large numbers of dispersing Colorado potato beetles in the spring and fall. Cold storage is also considered a physical control and, although it does not necessarily kill the insect pests, it at least stops their development and further feeding on the stored crop. Other methods include hand picking of pests, sticky boards or tapes for control of flying insects in greenhouses and various trapping techniques.

Chemical Control

If all other integrated pest management tactics are unable to keep an insect pest population below an economic threshold, then use of an insecticide to control the pest and prevent economic loss is justified. In most cropping systems, insecticides are still the principal means of controlling pests once the economic threshold has been reached. They can be relatively cheap and are easy to apply, fast-acting, and in most instances can be relied on to control the pest(s). Because insecticides can be formulated as liquids, powders, aerosols, dusts, granules, baits, and slow-release forms, they are very versatile.

Insecticides are classified in several ways, and it is important to be familiar with these classifications so that the choice of an insecticide is based on more than simply how well it controls the pest. When classified by method of application (site of encounter by insect), insecticides are referred to as stomach poisons (those that must be ingested), contact poisons, or fumigants. The most precise method of classifying insecticides is by their active ingredient (toxicant). According to this method the major classes of insecticides are the organophosphates, chlorinated hydrocarbons, carbamates, and pyrethroids. Others in this classification system include the biologicals (or microbials), botanicals, oils, and fumigants. Insecticides may be divided into two broad categories:

(a) conventional or chemical and

(b) biorational.

In this book we define conventional or chemical insecticides as those having a broad spectrum of activity and being more detrimental to natural enemies. In contrast, insecticides that are more selective because they are most effective against insects with certain feeding habits, at certain life stages, or within certain taxonomic groups, are referred to as "biorational" pesticides. These are also known as "least toxic" pesticides.

Because the biorationals are generally less toxic and more selective, they are generally less harmful to natural enemies and the environment. Biorational insecticides include the microbial-based insecticides such as the Bacillus thuringiensis products, chemicals such as pheromones that modify insect behavior, insect growth regulators, and insecticidal soaps. The majority of insecticides fall into the chemical category because they are typically more effective and can usually be used to control several pests. This provides the economic justification needed for the research and development of such products. In contrast, the more specialized market of the biorationals makes their long-term economic return less favorable.

Despite the advantages of conventional insecticides, the problems associated with their use have been well documented. These include the resurgence of pest populations after decimation of the natural enemies, development of insecticide-resistant populations, and negative impacts on nontarget organisms within and outside the crop system. One of the more serious problems is the development of resistance. Many insect pest species now possess resistance to some or several types of insecticides, and few chemical control options exist for these pests.

If all other integrated pest management tactics are unable to keep an insect pest population below an economic threshold, then use of an insecticide to control the pest and prevent economic loss is justified. In most cropping systems, insecticides are still the principal means of controlling pests once the economic threshold has been reached. They can be relatively cheap and are easy to apply, fast-acting, and in most instances can be relied on to control the pest(s). Because insecticides can be formulated as liquids, powders, aerosols, dusts, granules, baits, and slow-release forms, they are very versatile.

To compare pesticide use patterns, the dose, formulation or percent active ingredient of the product, and the frequency of application must also be considered and are used to calculate the EIQ Field Use Rating. Pesticides with a lower EIQ Field Use Rating have a lower environmental impact. The EIQ concept and techniques are evolving, but this, or a smilar rating, will provide information for appropriate pest management decisions in the future.

Natural enemies are generally more adversely affected by chemical insecticides than the target pest. Because predators and parasitoids must search for their prey, they generally are very mobile and spend a considerable amount of time moving across plant tissue. This increases the likelihood that they will contact the insecticide. When an insecticide is applied, ideally only the target pest(s) should be affected. The goal is to maximize pest mortality while minimizing harm to natural enemies.

The following factors, some of which are used when determining the EIQ, influence the impact of insecticide applications on natural enemies:

1. **Spectrum of activity**. Insecticides toxic to most insects (broad-spectrum) will more adversely affect natural enemies than materials that are narrow-spectrum or selective for specific insect species or life stages. Most insecticides in use today have a broad spectrum of activity.

2. **Residual (half-life) activity of insecticide**. Insecticides that remain toxic to pests for a long time and remain on the treated surface will have a similar effect on natural enemies.

3. **Coverage and formulation of insecticide**. Full coverage sprays will generally have a greater impact on natural enemies than directed sprays, systemics, or bait formulations. Spot or edge treatments directed at localized pest infestations or to a specific plant surface

most often occupied by the pest will be less detrimental than those applied to the entire field or plant.

4. **Dosage and frequency of application**. Higher rates and repeated applications will have a more detrimental impact on natural enemies.
5. **Timing of application**. Applying insecticides when natural enemies are not abundant or are less susceptible, such as when immatures are encased in host eggs, can be helpful.
6. **Susceptibility of natural enemy to insecticide**. Some natural enemies are inherently more resistant to insecticides than others, and some populations of natural enemies have been selected either naturally or through the efforts of researchers to possess higher levels of resistance. For example, strains of predaceous mites, rove beetles, lacewings, and species of parasitoids have been selected for increased levels of resistance to insecticides, thereby increasing the likelihood that some will survive insecticide treatments directed at pests. Some of these pesticide-tolerant strains are commercially available. Applications of some fungicides directed at controlling plant diseases can also reduce the incidence of fungal diseases of insects. In general, little can be done about this incompatibility except to use the fungicide only as needed.

In addition to killing natural enemies directly, insecticides may also have sublethal effects on insect behavior, reproductive capabilities, egg hatch, rate of development, feeding rate, and life span. Fungicides and herbicides may also have lethal and sublethal effects on natural enemies.

3

Parasitoids - Major Groups of Natural Enemies

Generally, arthropods that eat other arthropods are considered to be either predaceous or parasitic. Predators, which have been discussed in previous issues of this series, are characterized as follows:

- they are usually larger than their prey;
- they may be predaceous in either the immature stage, the adult stage, or both;
- they eat many prey during the course of their life;
- they are generally fairly mobile so that they can find or catch their prey; and
- most species tend to have a fairly broad host range.

Those insects that parasitize other insects have biological characteristics quite different than predators:

- they are slightly to substantially smaller than their host insect;
- all have a complex immature development that includes larval and pupal stages and it is only the larval stage that is parasitic;
- each larva kills only one host during its development;
- the larvae are totally restricted to parasitizing the single host insect and are often found internally in the host and therefore are not mobile in the environment; however, the adult stage is usually winged and mobile, and seeks the appropriate environment or host for its offspring; and

- they tend to be more host-specific than predators.

The term "parasite" is frequently used for insects that parasitize other insects. Many biological control workers, however, point to significant biological differences between such insects and other types of parasitic animals. "True" parasites are usually much smaller than their host, have a shorter life cycle than their host, and usually do not kill their host; examples include tapeworms and ticks. Recognizing this distinction, Reuter in 1913 coined the term "parasitoid" for insects that parasitize other insects. Some biological control workers prefer to use "parasite" whereas other use "parasitoid", so both terms can be found in MBCN.

Most agricultural and forestry pests are attacked by one or more parasitoid species. It is estimated that, on a world-wide basis, there are about 68,000 species of parasitoids that are known to science and have been given a scientific name. This constitutes a little under 10% of all known insect species. Entomologists generally believe that only about 10% of all insect species are known to science, and that possibly as many as 800,000 species of parasitoids actually exist. Most parasitoids are found within two major groups of insects, the orders Diptera (the true flies and their relatives), with about 15,000 known species of parasitoids, and Hymenoptera (bees, wasps, and ants), with about 50,000 known species. Within the Diptera, one family in particular the Tachinidae (tachinid flies), is especially important in the natural and biological control of agricultural and forestry pests. But, by far, the Hymenoptera comprise the largest group of parasitoids, and constitutes the overall most important group of natural enemies used in the biological control of insects.

However, the parasitoid way of life exists in several other groups as well, and an estimated 3,000 species exist in the other major groups of insects, most of these within the order Coleoptera (the beetles). The following families of beetles contain parasitoids:

Ground beetles—Family Carabidae

The carabids are one of the largest orders of beetles, with about 30,000 known species. Most are predators and many are important in natural and biological control. About 500 species of ground beetles are parasitoids, primarily of soil-dwelling insects. These are ectoparasites, feeding attached to the outside of their single host. Many of these are in the genus *Lebia*, which includes parasites of members of the leaf beetle family (Chrysomelidae). *Lebia grandis* parasitizes Colorado potato beetle pupae.

Rove Beetles—Family Staphylinidae

This is another large family, with about 30,000 species worldwide. Many are predators and some of these are important in biological control; about

500 species are known to be parasitoids. Most of the parasitoids attack the pupae of flies, such as filth-breeding flies (house flies, blow flies, flesh flies, etc.). One species, *Aleochara bilineata*, is an important parasite of cabbage maggot.

Family Rhipiphoridae

This is a small family of about 400 species worldwide. Most are parasites of the larvae of bees and wasps, and are not of importance in biological control. A few species of *Rhipidius* are parasites of cockroaches. The wingless, larviform female lays eggs on the ground in areas frequented by the hosts. The first instar attaches itself to a passing cockroach and enters the body, completing its larval feeding as an endoparasite but emerging to pupate.

Family Rhipiceridae

The cedar beetles comprise a very small family, with only about a half dozen species known in the United States. Members of the genus *Sandalus* parasitize the nymphs of cicadas in the soil, but are not very abundant.

Blister beetles—Family Meloidae

The larvae of some blister beetles attack ground-nesting bees; others attack the egg cases of grasshoppers. Because egg cases contain many eggs, some people consider this group of blister beetles to be predators. But because each larva consumes only a single egg case, others consider these to be egg-case parasitoids.

The order **Strepsiptera**, or twistedwing insects, is a small group of uncommon parasites closely related to (and sometimes included in) the beetles. Bees, wasps, leafhoppers, and planthoppers are the most common hosts, but a few species attack grasshoppers or bristletails. None are important in biological control.

The order Neuroptera consists of the lacewings, antlions, and their relatives. Most neuropterans are predators; one family consists of parasitoids.

Mantidflies—Family Mantispidae

Adult mantidflies are predators and have the front legs strongly modified for catching and holding prey, similar to praying mantids. The larvae are all parasitic within the egg sacs of spiders, especially ground-dwelling spiders. Mantidflies are more common in southern states.

The true flies (insect order Diptera) are characterized by having only one pair of wings that are functional for flight, with the hind pair of wings greatly reduced to tiny, knob-like structures, called halteres, that are involved in balance and orientation. Although the Diptera is a very large and important group of insects, relatively few families of flies are parasitoids. However, all of the species of one large family, the tachinid flies, are parasitic, and many species parasitize important agricultural pests.

Bee Flies—Family Bombyliidae

Many of these flies are stout with densely hairy bodies and a long, slender proboscis which is used for extracting nectar from flowers; their overall appearance and behavior gives rise to the common name for the family. The adults frequently hover over or rest on flowers or the ground in sunny places. When at rest, they tend to hold their wings outstretched, perpendicular to the body. This is a large and widely distributed group, with about 800 species. They occur throughout the United States, but are more common in the arid areas of the Southwest. The larvae are parasitic on cutworms, beetle grubs, bee and wasp larvae (including some parasitoids), or grasshopper egg pods. Those that attack grasshopper eggs are sometimes considered predators rather than parasites because several host individuals (in the egg stage) are killed. But functionally, the entire egg pod can be considered as a single host that is parasitized by the bee fly larva. In the few studies that have been conducted, the degree of parasitism of egg pods can be in the range of 50-80%, but in other locations, there is much less bee fly activity. Overall the family is considered beneficial, even though some species may be detrimental because they parasitize predators, other parasitoids, or pollinator bees. The few attempts to use bee flies in classical (importation) biological control of grasshoppers have been unsuccessful.

Tangle Veined Flies—Family Nemestrinidae

This family is related to the bee flies. There are only about six North American species, and most occur in the western United States. In most locations, these are rather rare flies, but they can be locally common, especially where appropriate hosts are abundant. The adult flies are of medium size and are stout-bodied; some are noticeably hairy and with a bee-like appearance. They are often seen hovering in open fields, or taking nectar from flowers. They primarily parasitize grasshoppers, but some are known to be parasites of the soil-dwelling grubs of scarab beetles.

Bigheaded Flies—Family Pipunculidae

These tiny, usually black flies have a very large head covered by huge eyes. The larvae are parasites of various species of leafhoppers and planthoppers. Although not uncommon, they are rarely abundant, and their role in biological control of pest leafhoppers is probably limited. There are about 100 species.

Thickheaded Flies—Family Conopidae

These are medium sized flies, with the head wider than the thorax, hence the common name. They are usually brownish in color, and often the base of the abdomen is slender, so they somewhat resemble wasps. The adults are often seen on flowers. This group, with about 70 species, is widely distributed in the United States and Canada, but nowhere are they abundant. Although there are some records of larvae parasitizing grasshoppers, some species are considered to be detrimental because they parasitize important pollinators, such as honey bees and bumble bees, but generally they are not sufficiently common to seriously reduce the populations of pollinators. This family has not been used in biological control.

Flesh Flies—Family Sarcophagidae

This is a large and common family, with well over 300 species in North America. Flesh flies are large grey, black, and white flies, related to blow flies. Many are scavengers of animal matter, with the larvae feeding on carrion, manure or dead insects, but a good proportion of the family are parasitoids of other insects, especially beetles, grasshoppers, or caterpillars. Some parasitoid species are sufficiently common to be considered important for biological control. For example, *Sarcophaga aldrichi* is an important parasite of the forest tent caterpillar, an important forest pest in much of the U.S. and southern Canada. In some locations, especially in the area of the Great Lakes, forest tent caterpillar goes through cyclic outbreaks causing considerable defoliation to hardwood forests. *S. aldrichi* responds to these outbreaks, building up in large numbers due to the abundance of hosts, and ultimately the tent caterpillar population crashes. The adult flies, which are large, droning, active fliers, can become incredibly abundant. Because they are attracted to sugars, they can be a considerable nuisance.

Tachinid Flies—Family Tachinidae

All species of this family are parasitic on insects and other arthropods. Some scientists consider this to be the largest family of flies, with about 1300 species known from the U.S. and Canada. Many are very important natural enemies of a variety of plant pests, especially caterpillars, but also beetles, grasshoppers, and sawflies. The tachinid flies have already been reviewed in this newsletter.

The order Hymenoptera includes the sawflies, bees, wasps, and ants. The term "wasp" usually conjures up a negative image by the lay person because of such creatures as hornets and yellowjackets that can sting, and can be somewhat aggressive when they or their nests are disturbed or threatened. But the group known as wasps by entomologists is much larger in scope, and the vast majority of wasp species are incapable of stinging.

Many of these stingless wasps appear to have an external stinger, which, however, is the egg-laying structure (ovipositor). In the stinging bees and wasps this organ is no longer used as an ovipositor, but has become modified for defense.

The stingless wasps comprise the single largest and most important group of natural enemies of insect pests. It has been said that every species of insect that attacks plants is itself attacked by one (or usually more) species of parasitic wasp. I'm not sure that this statement has ever been validated, but at least its essence is correct. There are about 18,000 species of Hymenoptera; the great majority of the these are wasps that parasitize other insects. Parasitic wasps are very diverse in their size, biologies, life cycles, and types of hosts attacked. The smallest known insects, less than 0.5 mm, are the fairyflies in the family Mymaridae, which spend their entire parasitic life at the expense of the egg stage of another insect. At the opposite extreme, some of the wasps that parasitize large grubs or wood borers may be more than 3 inches long.

Because of their abundance and importance, we can not cover the entire group in one article; therefore I will spread the information out in a continuation of this series. As in past articles, I will take a taxonomic approach. The parasitic wasps fall into several superfamilies, which are then subdivided into families; these will be the taxonomic groupings that I will use. In this article, we will start by discussing the wasps in the superfamily Ichneumonoidea.

The Ichneumonoidea is a very large and important group; all of the species are parasitoids. There are two major families, the Braconidae and Ichneumonidae; both are very important in natural and biological control.

Braconid wasps comprise a large group of over 1700 North American species; all are parasites of insects or other arthropods. Most are small wasps, ranging in size from 2-3 mm up to about 15 mm (1/2 inch) in length. They are generally black or brown in color, but some have yellow, orange, or red markings. They have long and noticeable antennae, and the ovipositor may also be long and readily apparent.

Caterpillars constitute the largest group attacked by braconids. Other common hosts include flies, sawflies, wood boring beetles, weevils, leafmining insects, true bugs, and ants. One important subfamily, the Aphidiinae (sometimes considered to be a separate family, the Aphidiidae) are very small wasps that parasitize aphids. Braconids usually parasitize the immature stages (larvae, pupae, or nymphs) of their hosts. A few parasitize adult insects, such as *Microctonus aethipoides,* which parasitizes and sterilizes adult alfalfa weevils.

The generation time of most braconids is relatively short, often in the range of 10-30 days. Many braconid larvae leave their host just prior to pupation, which usually occurs within a silken cocoon. Some species produce only one parasite per host insect; others may have several-even a dozen or more-if the size of the host can support them all.

Braconids are important in the natural and biological control of many serious pests of the farm, forest, and garden. Several exotic species have been introduced and permanently established for the control of alien pests. A few species are available commercially for augmentation biological control.

Some Important Species of Braconid Wasps

Species	*Host*
Aphaereta pallipes	root maggots
Aphidius spp.	aphids
Bracon hebetor	moth larvae in stored grain
Cotesia glomerata	imported cabbageworm
Cotesia melanoscela	gypsy moth
Dacnusa sibirica	leafminers
Diaeretiella rapae	aphids
Lysiphlebus testiceipes	aphids
Macrocentrus ancylivorus	leafrollers/caterpillars
Macrocentrus grandii	European corn borer
Microctonus aethiopoides	adult alfalfa weevil
Microctonus vittatae	striped flea beetle
Microplitis croceipes	corn earworm
Peristenus digoneutis	tarnished plant bug
Pholetesor ornigis	spotted tentiform leafminer

Ichneumonid wasps, or ichneumons, constitute one of the largest groups of insects, with over 3100 species in North America. Adults vary a lot in size, form and color, but generally they are larger than braconids. Size is somewhat host-dependent, and some that parasitize large hosts may be 1½ inches long. The ovipositor is often quite long, sometimes longer than the body. As with the braconids, the antennae are long and noticeable. Most are black, brown, or tan, but some are more brightly colored.

Many types of insects are attacked by ichneumons, though most individual species of wasp attack a relatively narrow range of hosts. Common types of hosts include caterpillars and the larvae of beetles and sawflies.

The larval stage of ichneumons varies from about 10 days to several weeks. Many species have a single generation per year but others have two, three, or more generations annually.

Although ichneumons are very important in the natural control of many plant pests, surprisingly few have been actively used in biological control programs.

Some Important Species of Ichneumonid Wasps

Species	*Host*
Bathyplectes spp.	alfalfa weevil
Collyria coxator	wheat stem sawfly
Diadegma insulare	diamondback moth
Eriborus terebrans	European corn borer
Lathrolestes nigricollis	birch leafminer
Scambus applopapi	nantucket pine tip moth

The chalcid wasps (Superfamily Chalcidoidea) comprise a large group with over 2000 North American species. Nearly all species are very small; some are the smallest of all known insects. There are a great number of pests attacked by members of this group and different parasite species attack different host stages from egg to adult. Because of their small size, the wing venation is greatly reduced and generally distinctive. Also because of their small size, their presence and importance are often overlooked, but they are considered to be one of the most important groups in both natural control and applied biological control.

Two interesting biological trends occur within this group. First, there is a tendency toward plant-feeding, even in the larval stage, and some species are agricultural pests (even though others in the same family or even genus are insect-parasitic). One family known as the fig wasps are responsible for pollinating figs, and their larvae feed within the developing fig fruits. Second, there is a tendency toward hyperparasitism. Hyperparasitism is the parasitism of another parasite larva that is already associated with the original host. For example, one species, called the primary parasite, may parasitize a caterpillar. The hyperparasite parasitizes the primary parasite. Sometimes this is a facultative association, where the hyperparasite may attack either the original host or the primary parasite. In many cases, hyperparasitism is obligatory. In extreme cases, called autoparasitism, male parasite larvae parasitize female larvae of their own species.

Some taxonomic relationships in the Chalcidoidea are unclear, and therefore the classification of the superfamily is unsettled. About 20 North American families are recognized. Of these, the following six are very important in natural and biological control.

Fairyflies—Family Mymaridae

Some of the fairyflies are the tiniest of all insects, being 0.5-1 mm in length. The smallest ones known are about 0.2 mm (0.008 inch!). They are distinctive because the wings (at least the hind wings) are edged by a fringe of long hairs, but a microscope is needed to see this. All known species are parasites of the eggs of other insects. Common host groups include the flies, beetles, booklice, and leafhoppers. Many species important in biological control are in the genus *Anaphes*. *A. flavipes* is an introduced parasitoid of the eggs of cereal leaf beetle, a serious pest of small grains. *A. iole* is being investigated as a natural enemy of tarnished plant bug. *A. luna* contributes to the overall biological control of alfalfa weevil. *Anagrus* is another important genus. For example, *A. epos* parasitizes the eggs of white apple leafhopper and *A. nigriventris* parasitizes the eggs of potato leafhopper.

Trichogramma wasps—Family Trichogrammatidae

Like the fairyflies, the trichogrammatid wasps all are parasites of the eggs of other arthropods and therefore all are very small, mostly less than 1 mm in size. They have been used much more often in biological control than have fairyflies, and constitute the single most widely used genus of arthropods in augmentation biological control programs. Several species are commercially available. Common species in the United States include *Trichogramma minutum*, *T. platneri*, and *T. pretiosum*, but there are about 40 other species native to the U.S. and Canada. Some foreign species are being evaluated for augmentation biological control, such as *T. ostriniae*, *T. nubilale*, and *T. brassicae* for control of European corn borer. Many *Trichogramma* species have a wide host range and will attack the eggs of many species and even many families of insect hosts. Common hosts are the eggs of moths and butterflies, beetles, flies, wasps, and true bugs. Several species are aquatic, attacking the eggs of aquatic insects such as diving beetles and dragonflies. Some aquatic trichogrammatids use their wings as oars, to swim through the water in search of host eggs.

EULOPHIDS—FAMILY EULOPHIDAE

This is one of the largest families of the Chalcidoidea, with about 3400 species worldwide and more than 600 in North America. The wasps themselves are rather small (1-3 mm), and often are a brilliant metallic color (commonly metallic blue or green). This is a biologically diverse group, with some species attacking mites (as predators), spider egg cases, scale insects, and thrips, but most attack beetle or moth larvae or pupae, flies or other wasps. Many parasitize leaf-mining and wood-boring insects. Many major crop pests are attacked by eulophids. The following are some examples from here in the Midwest. *Sympiesis marylandensis* is an important parasite of spotted tentiform leafminer of apple. *Diglyphus isaea* is an important parasite of agromyzid leafminers and is available from commercial suppliers of natural enemies; it is used primarily in commercial greenhouses. *Edovum puttleri* is a parasite of the eggs of Colorado potato beetle. *Pediobius foveolatus* is commercially available and parasitizes larvae of Mexican bean beetle; it was originally introduced from India. *Tetrastichus* is a genus with several parasites of plant pests; for example, *T. asparagi* parasitizes the asparagus beetle. Sometimes several species of eulophid will attack one type of host. For example, *Elachertus cacoeciae, Horismenus microgaster, Pediobius crassicornis, Pediobius facialis,* and *Sympiesis bimaculatipennis* are all parasitoids of obliquebanded leafroller, a fruit tree pest. *Ceranisus* sp. is being investigated as a biocontrol agent of western flower thrips larvae.

Pteromalid wasps—Family Pteromalidae

This is another large family, with about 350 known species in the U.S. and Canada. It is a very diverse group and many types of insects serve as hosts to wasps in this family. Common types of hosts include the larvae of moths, flies, beetles, and wasps. A few parasitize scale insects and mealybugs. Many are hyperparasitoids, parasitizing other parasitoids within a variety of hosts. Several have been used in classical biological control. The following are some examples of pteromalids important in natural and biological control in the upper Midwest. *Pteromalus puparum* is a parasite of the pupae of the imported cabbage worm. This is a gregarious species, and as many as 100 or more parasites can develop from one host pupa. *Anisopteromalus calandrae* is a parasite of the larvae of beetles that infest stored grain; this species is commercially available for release in grain storage and handling facilities. Several types of pteromalids parasitize the pupae of filth-breeding flies, such as flesh flies, house flies, blow flies, and stable flies. Examples include parasites in the genera *Nasonia, Muscidifurax,* and *Spalangia*. Some species occur naturally in our area, and some species are commercially available for release on dairy farms or at cattle feedlots.

Encyrtid wasps—Family Encyrtidae

This family, with over 3000 known species in the world, and about 400 in the U.S. and Canada, is one of the most important chalcidoid families for biological control. Again, there is a great diversity in types of hosts. Many are parasites of soft scales, armored scales, and mealybugs, and many of the very successful cases of classical biological control of scale and mealybug pests of fruit trees have involved encyrtid wasps. Other hosts include the eggs or larvae of insects in about 15 families of beetles, 10 families of flies, and 20 families of moths and butterflies. Less common hosts include eggs, nymphs, or adults of the true bugs (such as stink bugs), katydid eggs, and the nymphs of certain ticks. A couple important examples of encyrtids include *Ooencyrtus kuwanae*, an introduced gypsy moth egg parasite (see MBCN Vol. IV, No. 6) and *Copidosoma floridanum*, a native parasite of cabbage looper larvae. Several species of encyrtids are commercially available for control of scale insects and mealybugs in greenhouse, conservatories, and interior plantscapes. Two examples are *Leptomastix dactylopii* which is a specialized parasite of citrus mealybug, and *Metaphycus helvolus* which is a general parasite of many species of soft scale, including black, hemispherical, brown soft, nigra, and citricola scales. As I sit working at my desk, it is not at all unusual to see a tiny parasitic wasp walking or hopping across my papers—*Comperia merceti*, an egg parasite of cockroaches. Our building has a permanent infestation of the German cockroach, and a naturalized population of this little parasite contributes to the reduction of cockroach populations.

Aphelinid wasps—Family Aphelinidae

Because of its importance in the control of scale insects, this family is considered one of the most important in all of biological control. However, there is confusion over the proper classification of this group, and in various references it may be included within the Encyrtidae or Eulophidae. One reason for the taxonomic difficulties is the very small size of these insects; most are under 1 mm. The aphelinids are most commonly associated with various families of Homoptera, including scale insects, mealybugs, whiteflies, and aphids. Some are egg parasites and additional hosts occur in various insect groups. One of the most important members of this group is *Encarsia formosa*, a parasite of greenhouse whitefly that is commercially produced and widely used. Members of the genus *Eretmocerus* are also whitefly parasites and are important in the control of silverleaf whitefly. Various *Aphelinus* species are aphid parasites, such as *A. varipes*, a parasite of greenbug, and *A. mali*, a parasite of woolly apple aphid. A complex of foreign *Aphelinus* species have been introduced into the U.S. for control of Russian wheat aphid.

The above six families of the Chalcidoidea are the most important in biological control. The following summarizes the remaining families.

The small family **Leucospidae** has only a half dozen species in the U.S. and Canada. Relatively little is known about the biology of this family, but all known species are parasites of the stinging bees and wasps. Members of the family Chalcididae can be recognized by their enlarged hind legs. There are somewhat over 100 species in the U.S. and Canada. Most attack the pupae of flies or moths. *Brachymeria intermedia* is a parasite of the pupae of gypsy moth.

The family **Eurytomidae** is of intermediate size, with about 250 known species in North America north of Mexico. Some are plant feeders including gall-makers and seed-feeders. There are a few important agricultural pests, such as the alfalfa seed chalcid. The parasitic species of eurytomids largely attack host insects found within plant tissues, such as stems, seeds, or galls. There has been little if any usage of eurytomids in biological control of insects, but some of the seed-feeders have been used in biological control of weeds.

The family **Torymidae**, with about 170 species in the U.S. and Canada, is biologically similar to the eurytomids in that many are plant feeders while others are parasitic on other insects. At least one species, the apple seed chalcid, can be an agricultural pest. Torymid wasps are often brightly colored, especially metallic blue or green. Those that lay their eggs in plant tissue have very long ovipositors. Many species are parasitic on gall insects. Other hosts include the eggs and larvae of moths, beetles, and flies. Praying mantid eggs are often parasitized by torymids.

The family **Eupelmidae** is of moderate size, with about 100 species in North America. Many eupelmids attack hosts within plant tissues, such as gall-makers or the larvae of wood-boring beetles. Other eupelmids are egg parasites. *Anastatus disparis,* an egg parasite of gypsy moth, was introduced from Europe into the U.S. for biological control and is now established here.

All members of the fig wasp family, **Agaonidae**, are pollinators of figs, with the larvae feeding within the fruit receptacle; there are very few species in North America. Another very small family is the **Ormyridae**, with all known North American species parasitic on the larvae of gall wasps on oak trees. All members of the small family **Eucharitidae** are parasitic on the larvae and pupae of ants. The small family **Perilampidae** consists mostly of hyperparasites, parasitizing the larvae of fly and wasp parasites of primarily moth hosts; a few species are primary parasites of beetle and lacewing larvae. The very small family **Tanaostigmatidae** are all believed to be gall-formers. The small family **Signiphoridae** consists mostly of hyperparasites, although some are primary parasites of scale insects. Another small family, **Elasmidae**, also consists of both primary and hyperparasites.

In summary, the superfamily Chalcidoidea is quite large and of diverse biological habits. Although there are many families, relatively few contribute significantly to natural and biological control of pests. However, the contributions from these important families has resulted in many noteworthy successes in biological control.

There are three families in the superfamily Cynipoidea. The largest family, with over 600 species in the U.S. and Canada, is the **Cynipidae**, which is commonly called the gall wasp family. Gall wasps are generally quite small, usually less than a few millimeters in length. These insects are plant-feeders in the larval stage, causing abnormal growths, called galls, on leaves or stems; the larvae feed within these galls. Although these insects are not themselves parasitic, they account for some very biologically-interesting relationships that include parasitoids. The galls created by cynipids (such as on oak trees) are frequently inhabited by other wasp larvae. Some of these are just taking advantage of the resources provided by the gall, and they co-exist with the larval cynipid; these are called inquilines. However, both the gall wasps and their inquilines are frequently parasitized by other wasp species, which, in turn, may also be parasitized by yet other species. It is not unusual for four, five, or even more species of these tiny wasps to emerge from a large oak gall created by just a single species of gall wasp.

Two other groups are very closely related to the gall wasps; some people consider these to be in sister families, while other scientists contend that they are closely enough related to be considered subfamilies of the Cynipidae. The **eucoilid** wasps, of which there are about 80 species in our part of the world, are mostly all parasites of the pupae of various types of flies. Some species are relatively common parasites of fly maggots that occur in manure and carrion; at least one species has been established in the southwestern United States for classical biological control of such pests. There are only about 30 species of **charipid** wasps in the U.S.; these are all hyperparasitoids; that is they parasitize other parasitoids usually the braconids and chalcids that parasitize aphids.

Also within the Cynipoidea are the two small families **Figitidae** and **Ibaliidae**. Figitid wasps parasitize the pupae of lacewings and the larvae and pupae of flies. Some species parasitize pest flies (such as dung-breeding species) while others parasitize beneficial flies, such as predatory hover (syrphid) flies. Ibaliid wasps are the largest species in the superfamily, with some over a half inch long. There are only a few species, and generally not common. They parasitize the larvae of horntails, which are woodborers.

Members of the small (less than 60 known species in the U.S. and Canada) superfamily Evanioidea are all parasitoids of other insects. Members of the family **Evaniidae** are all parasitoids of cockroach eggs; some species are fairly common in human habitations that are infested with cockroaches. **Gasteruptiid** wasps are parasitoids of the natural enemies of solitary bees and wasps. **Aulacid** wasps are parasites of the larvae of wood-boring insects.

The superfamily Pelecinoidea is represented by a single species in North America, in the family Pelecinidae. *Pelecinus polyturator* is a large, spectacular parasite of the white grub larvae of June beetles (*Phyllophaga* spp.). It is a relatively common species through much of the Midwest. Females are jet black, two inches long, with a very long, slender abdomen. Males are very rare.

The Proctotrupoidea is a fairly large superfamily; all known members are parasitoids. There are about eight families in total, but some of these have only one or two species and are quite rare and therefore will not be covered. The families **Proctotrupidae** (also known as the **Serphidae** not SYRPHIDAE, which is a family of flies), **Ceraphronidae**, and **Megaspilidae** are all fairly small, but members are non uncommonly found during general insect collecting. Not much is known of their biologies, but some are parasites of the larvae of beetles, flies, or other hosts; some are hyperparasites. **Diapriid** wasps are fairly common, with over 300 species known; most are parasites of fly larvae. From a biological control perspective, the families **Platygastridae** and **Scelionidae** are the most important in this group. Platygastrid wasps are also fairly common; they parasitize gall midges, mealybugs, and whiteflies. *Platygaster hiemalis* is an important native parasite of the introduced wheat pest Hessian fly. Three species of *Allotropa* (two native and one intentionally introduced) are important parasites of Comstock mealybug in the eastern United States. All scelionid wasps are parasitoids of the eggs of other insects and spiders. The eggs of grasshoppers, praying mantids, and stink bugs are commonly parasitized.

Most of the wasps in the superfamily Bethyloidea (or Chrysidoidea) are parasitoids of other insects. The family **Chrysididae** includes the cuckoo wasps. These are medium-sized, stout-bodied insects, somewhat resembling the shape of small bees. They are usually metallic blue, green, or red in color and they curl up into a ball when disturbed (kind of like tiny winged gazing balls!). Some are inquilines in bee or wasp nests but others parasitize sawfly larvae or the eggs of walkingsticks. **Bethylid** wasps, of which there are about 100 species in the U.S., are primarily parasitoids of the larvae of moths and beetles. Several species are important in the biological control of forest pests such as hickory shuckworm and Nantucket pine tip moth. Carpenter ants are also parasitized by bethylids. *Laelius pedatus* has been researched for augmentative biological control of pests of stored products, but I have not yet seen that it has become commercially available. Members of the family **Dryinidae** are primarily parasites of nymphs and adult leafhoppers, planthoppers, and treehoppers, but they are not at all common and probably have little if any benefit in pest management.

Some of the wasps in the superfamily Scolioidea are amongst the largest of all parasitoids. Contrary to most groups of parasitic wasps, females of this group are capable of using their ovipositor as a stinger for defensive purposes. Some, such as members of the velvet ant family (**Mutillidae**) are notorious for their highly painful stings. Velvet ants are densely clothed in often brightly colored hairs (yellow, orange, red, or white, often with black bands) which serve as warning coloration. Females are wingless. They parasitize the larvae of ground-nesting bees and wasps. Closely related to the velvet ants is the family **Tiphiidae**, whose members are usually brown or black and sparsely hairy. Tiphiid wasps parasitize larval insects found in the soil and in rotting wood; many are parasites of serious pests such as white grubs. *Tiphia intermedia* is a common species that parasitizes June beetle (*Phyllophaga*) larvae. *Tiphia vernalis* and *T. popilliavora* were previously discussed in MBCN as parasitoids of Japanese beetle grubs. Many **scoliid** wasps are quite large, some up to two inches in length. They are often brightly marked with yellow and black warning coloration. These, too, are parasitoids of white grubs; many attacking the larvae of June beetles.

WASPS

Anaphes flavipes (Hymenoptera: Mymaridae)

Anaphes flavipes is a parasitoid wasp that was introduced to the United States from Europe in 1965 for the control of the cereal leaf beetle, *Oulema melanopsis L.* (Coleoptera: Chrysomelidae).

Appearance

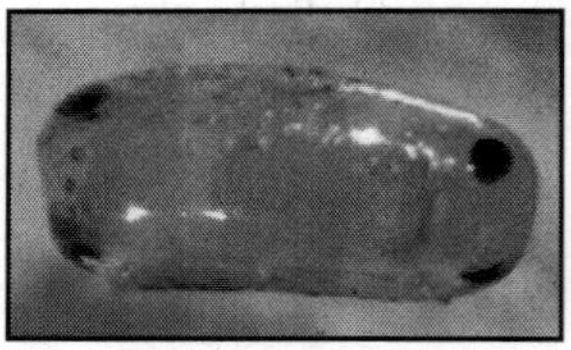

Anaphes flavipes

Anaphes flavipes is a minute (approximately .75 mm) chalcidoid wasp with greatly reduced venation in the forewings. The antennae are long and slender and are used to distinguish the sex of the wasp. The males are characterized by filiform, 12-segmented antennae, where the females have clavate 9-segmented antennae. The tarsi are 4 or 5 segmented. Females are generally more active than males, walking briskly with their antennae in alternating but constant tapping motions.

Habitat

Usually found in small grains fields or in adjacent hedges or fence rows. The wasps' overwinter habits are presently unknown.

Pests Attacked (Host Range)

Oulema melanopsis (cereal leaf beetle) is the main host; however *Anaphes* is also known to parasitize *O. gallaeciana, Lema collaris,* and *Lema trilineata* (three lined potato beetle) if the primary host is not available. Techniques have been developed using *L. trilineata* eggs for laboratory rearing of *A. flavipes*. With the exception of the four species listed previously, *Anaphes* will oviposit in eggs of other Chrysomelidae, but development doesn't proceed much beyond embryogenesis.

Life Cycle

Anaphes can have on average 2 generations per year, although as many as 8 have been observed on *Oulema*. Their life cycle consists of the egg, larvae, prepupa, pupa, and adult stages. The egg stage consists of the internal cleavage of the parasitoid and its growth into a larva. The larval stage consists of the consumption of the host's egg yolk. As a prepupa, the parasitoid is completely motionless; the excretion of the fecal matter signifies the end to this stage. In the early pupal stage, the red compound eyes are the first feature visible. The body begins to darken and faint adult characteristics become visible.

In the late larval stage the body is dark and the ovipositor is visible on the female. Just before emergence the head and legs show movement as the parasitoid chews its way through the egg chorion. Under ideal conditions (21-25°C) the adults emerge from the host egg in 10 to 11 days. A minimum of 174 hours (at 32.2°C) to a maximum of 1089 hours at (2.7°C) was determined for the development of *Anaphes* from egg to adult. Within one hour after emergence, adult female *Anaphes* will attack suitable hosts and commence egg deposition. On average, females deposit 20 eggs during an average 2-3 day post emergence period. Females will deposit fertilized (resulting in female offspring) and non-fertilized eggs (which result in male offspring).

Studies have indicated that when *Oulema* eggs are most abundant, cool temperatures are prevalent. These cool temperatures are not conducive to the rapid development of *Anaphes* and may be another explanation for the time lag between host and parasitoid activity.

Relative Effectiveness and Conservation

A. flavipes is one of two parasitoids that have been successfully established for cereal leaf beetle control. An equilibrium where the cereal leaf beetle was kept to controllable levels lasted until the late 1980's in Maryland when surveys found the natural enemies absent or with very low populations. This prompted a renewed release effort that is being continued today.

Alone, *A. flavipes* is usually unable to reduce host populations below an economically acceptable level. The original efforts of USDA that introduced a complex of parasitoids to control cereal leaf beetle was quite successful. This complex includes the larval parasitoids *Diaparsis temporalis, Lemophagus curtus,* and *Tetrastichus julis* .

The effectiveness of this parasitoid remains fairly low during peak cereal leaf beetle activity. This is attributable to a lack of synchronization between the parasitoid and its host. Measures to minimize this asynchronization are to avoid a heavy layer of ground litter as this may trap the parasites as they emerge. Also, a heavy cover may delay soil warm-up, attributing to the delay in parasite emergence.

It is suggested by Anderson that the *A. flavipes* may be sustained through the late summer and fall by eggs of *Oulema* that were deposited on marginal grasses. He noted a state of diapause and assuming this, he reasoned that one egg containing two females and one male would be sufficient to maintain the species and create a small focus of parasites on the margins of wheat fields in the spring.

Oulema eggs (at 21°C) are susceptible to parasitization up to 115 hours after deposition. Between 118-120 hours, *Anaphes* will lay eggs, but neither parasitoid nor host develops. The parasitoid will develop successfully in host eggs that are 115-118 hours old, but the time for development and emergence is retarded. If the development in the host egg has progressed to where the mandibular structures of the beetle embryo have sclerotized, the parasite will retract its ovipositor and not parasitize the egg.

Pesticide Susceptibility

Pesticide applications are detrimental to *A. flavipes* and thus to successful biological control of its host.

Atanycolus cappaerti (Hymenoptera: Braconidae)

The emerald ash borer (EAB) is a serious invasive insect pest that was first discovered in southwest Michigan in 2002 and affects ash species in North America. The high rate at which the emerald ash borer has been destroying ash trees has spurred research on the potential for biological control. In particular, researchers have been searching for parasitic wasps that could be raised in large numbers and released to provide suppression of EAB.

Biological control efforts have focused primarily on identifying and releasing egg and larval parasitoids imported from EAB's home range in Asia. The three most important natural enemies to date are the larval parasitoids *Tetrastichus planipennis* (Hymenoptera: Eulophidae) and *Spathius agrili* (Hymenoptera: Braconidae); along with an egg parasitoid named *Obius agrili* (Hymenoptera: Encyrtidae). In 2008 a native larval parasitoid *Atanycolus cappaerti* was discovered in Michigan. While it has been well documented that a species of the same genus (*A. hicoriae*) parasitizes the EAB, this newly described species is the first to attack at rates high enough to be considered significant for biological control. While studies are still ongoing, the use of the native parasitoid *A. cappaerti* for augmentative biological control would be desirable from an environmental perspective as it is already established and has a niche in the ecosystem.

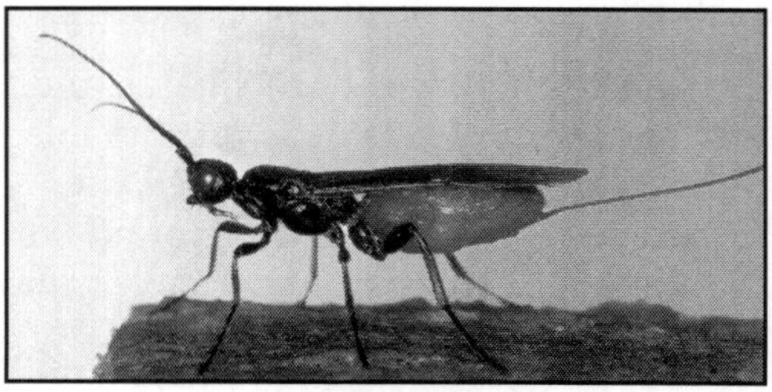

Atanycolus cappaerti

Appearance

While distinguishing A. *cappaerti* from A. *hicoriae* species is difficult, narrowing it down to one of these two species is easily done by sight. The female ranges from 5-7 mm and has a ovipositor that can be as long as 6 mm. The male is around 4 mm. The head and thorax in both species are jet black, while the abdomen is bright red. The wings are colored dark grey.

Habitat

Atanycolus cappaerti has so far only been identified at two sites in Michigan. Its status as a new species and the difficulty of distinguishing it from A. *hicoriae* mean its range is largely unknown.

Pests Attacked (Host Range)

Atanycolus cappaerti's full host range is not yet fully known. It has been documented parasitizing the larvae of two native beetles in the genus *Agrilus* in addition to EAB. These include the two-lined chestnut borer (*Agrilus bilineatus*) and the poplar borer (*Agrilus liragus*). These beetles can cause mortality in trees that are under stressed conditions like drought, but do not seem to attack healthy trees.

Life Cycle

Atanycolus cappaerti is a solitary ectoparasitoid. This means that the parasitoid lays only a single egg, which hatches into a larva that feeds externally on its host. The female adult uses her ovipositor to drill through the outer layer of bark and sap wood to parasitize wood boring larvae in their 2nd, 3rd, and 4th instars that are feeding on the phloem tissue of the tree (Figure 2); once the larvae move deeper into the wood to pupate, *A. cappaerti* can no longer parasitize them . Once a wasp identifies the larvae and drills down, she paralyzes it with a sting and deposits one egg on the outside of the host. This egg will hatch into a larva, which remains on the outside of its host and feeds (Figure 3) until it is ready to detach itself and pupate in the EAB gallery (Figure 4). This whole process takes around two weeks. In its cocoon *A. cappaerti* will pupate for about another 2 weeks and then emerge after chewing an exit hole through the bark, completing its life cycle after a total of approximately 30 days.

The wasp pupae overwinter in cocoons in borer galleries. They emerge as adults in May before EAB and parasitize the larvae that have overwintered in their galleries. These wasp larvae then develop and emerge around July to parasitize borer larvae that hatched from eggs laid in May. This cycle will repeat until October when temperatures decline and overwintering begins again.

Relative Effectiveness and Conservation

Research into the effectiveness of *A. cappaerti* is ongoing but promising. Parasitism rates have been highly variable, ranging from 9 -71% among study sites.

Pesticide Susceptibility

To date there has been no research on the effect of pesticides on *A. cappaerti*. It would be reasonable to assume that like most natural enemies it would be highly susceptible as an adult. The host gallery and cocoon may afford it some protection as a larva, from bark-applied insecticides, though any residual topical pesticide applied to the tree's bark may kill the newly pupated adult as it emerges. The main method of controlling EAB has been soil-applied and injected systemic insecticides. These chemicals are ingested by EAB as they eat the phloem tissue though which these pesticides are being transported. This could have serious consequences for *A. cappaerti* due to the death of its host or secondary consumption of the pesticide.

Bathyplectes anurus and *B. curculionis*
Hymenoptera: Ichneumonidae

Bathyplectes spp. are small, non-stinging wasps that are parasitoids of the alfalfa weevil, a serious pest of alfalfa in the Midwest and elsewhere. They were introduced to North America in 1911 from Italy by the U.S. Department of Agriculture as part of a biological control effort against the alfalfa weevil, also a non-native.

Appearance

Adults of both species are similar in appearance, about 3 mm long with black, robust bodies. Cocoons are about 3.5 mm long, brown, and football-shaped with a white band around the middle.

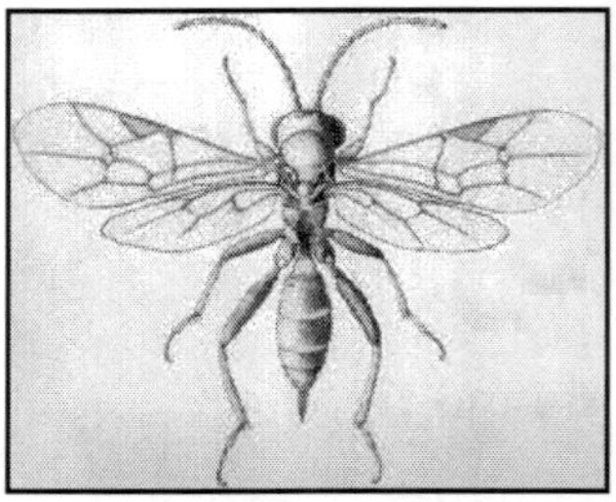

Bathyplectes anurus

The cocoons of the two species are the most visible sign of the parasitoid and also are the handiest for separating the species: in *B. anurus* the white band is raised, and the cocoon has the unusual habit of "jumping" when disturbed, whereas in *B. curculionis* the white band is not raised and the cocoons do not jump.

Habitat

Alfalfa fields.

Pests attacked

The alfalfa weevil, *Hypera postica,* and the Egyptian alfalfa weevil, *Hypera brunnipennis.*

Life cycle

The two *Bathyplectes* species are similar in habits. Both lay their eggs in alfalfa weevil larvae, preferring to oviposit in the early instars. The wasp larva that hatches from the egg feeds internally and slowly devours the weevil larva, ultimately killing its host after the weevil has finished spinning its cocoon. The parasitoid larva then emerges from the weevil and spins a cocoon of its own. Only one parasitoid can successfully develop in a host weevil.

To understand the life cycle of *Bathyplectes* spp., it is necessary first to understand the life cycle of the alfalfa weevil host.

The alfalfa weevil typically has just one generation per year, with larvae present during the spring. The alfalfa weevil adults emerge from pupae during late spring to early summer, feed for several weeks, and then spend the remainder of the summer "aestivating" in a state of arrested activity and development (diapause). Aestivation is completed by late summer or fall, and the adults become active while the weather remains favorable, then hibernate during the winter and resume feeding and laying eggs the following spring.

The adult flight activity of both *Bathyplectes* spp. is synchronized with the spring activity period of the alfalfa weevil larvae. The flight lasts up to several weeks, and peak parasitism levels occur 1 to 2 weeks prior to the peak in numbers of weevil larvae. *B. anurus,* which lays approximately 300 eggs, has just one generation a year, with all parasitoid pupae produced by spring parasitism undergoing diapause, and not emerging as adults until the following spring when weevil larvae are again abundant. *B. curculionis,* on the other hand, has a partial second generation; many of the parasitoid pupae from spring parasitism are in diapause, but some develop and emerge as adults which then must find and parasitize weevil larvae during the summer. *B. curculionis* females lay approximately 200 eggs per year.

Relative Effectiveness

Together with three other parasitoid wasp species introduced by USDA, the mymarid wasp, *Patasson luna,* which attacks weevil eggs; the eulophid wasp, *Tetrastichus incertus,* which attacks weevil larvae; the braconid wasp, *Microctonus aethiopoides* and *M. colesi,* which attack adults; and the fungal pathogen, *Zoophthora phytonomi*; *Bathyplectes* spp. keep alfalfa weevil populations in check most years in most locations. The egg predator,

Peridesmia discus, has recently been recovered, years after the initial releases were made. In areas where 3 or more parasites are established, alfalfa weevil is seldom of economic importance, even in areas once requiring multiple applications of insecticide.

In comparing the two *Bathyplectes*, *B. anurus* is considered the superior biological control agent, especially in hot climates. This is due in part to *B. anurus* having a 50% greater reproductive potential than does *B. curculionis*. Also weevil larvae are able in many cases to kill *B. curculionis* eggs through a process known as encapsulation. Where the two species occur together, *B. anurus* has tended to become dominant.

Conservation

Avoid the use of insecticides when adult parasitoids are active. Also, the cultural practice of leaving uncut "refuge" areas of alfalfa to avoid killing parasitized weevil larvae can be effective.

Catolaccus grandis (Burks) (Hymenoptera: Pteromalidae)

Catolaccus grandis is an ectoparasitoid of the boll weevil classified initially within the genus *Heterolaccus*. It was first introduced to the U.S. during the early 1970's and released in experimental fields in Mississippi State, MS. Researchers observed encouragingly high parasitism rates, but failure of this parasitoid to become established made it unsuitable for use within a classical biological control program. It was reintroduced a decade later and released in College Station, TX. Rearing methods for this parasitoid progressed quickly between 1985 and 1992 making *C. grandis* suitable to be used in a biological control by augmentation program.

Appearance

The female has a length between 4 to 5.5 mm. The head and thorax are shining black with the antennal pedicel and scape yellow or light tan, the funicle and club brown. The compound eyes are bright red. The legs have black coxa shading to tan, the trocanters and femora are tan and the tibia and basal tarsal segments are light yellow. The abdomen is red-brown with iridescent (metallic) blue or blue-green reflections. Males are smaller, 3.0-3.5 mm in length, and their abdomen differs in shape and coloration from that of females. The female abdomen is pointed metallic blue or blue-green, while the male abdomen is oval in shape and presents a transversal window white to light yellow in coloration.

Habitat

Catolaccus grandis originated in Southeastern Mexico and it occurs naturally in the Mexican states of Veracruz, Tabasco, Campeche, Yucatan,

Chiapas, Nayarit and Sinaloa. This parasitoid lives in the tropical and subtropical forest and parasitizes its natural host (*Anthonomus grandis*) in native host plants, which include *Hampea nutricia* Fryxell, *H. trilobata* Standley, *Cienfuegosia rosei* Fryxel, and also in wild and cultivated cotton, *Gossypium hirsutum* L.

Catolaccus grandis is a tropical insect which is not adapted to overwinter in temperate regions. In the states of Tabasco, Campeche, Veracruz, and Chiapas, this parasitoid inhabits the rain forest in clearings where the *Hampea* plants usually grow. Dryer climates are also suitable for this parasitoid. In the states of Yucatan, Nayarit, Chiapas, and Sinaloa, this parasitoid survives in wild or cultivated cotton and short brush where *Cienfuegosia* plants proliferate.

Pests Attacked

Only two species are known to be natural hosts of *C. grandis*: the boll weevil, *Anthonomus grandis* and a closely related species, *Anthonomus hunteri*, which inhabits the Yucatan peninsula. This parasitoid seems to be specific and highly adapted to its natural host; however, under laboratory conditions *C. grandis* has been successfully reared in three other species including the cow pea weevil, *Callosobruchus maculatus* (F.), the cow pea curculio *Chalcodermus aeneus* Boheman, and the pepper weevil *Anthonomus eugenii* Cano with significant reduction in fitness. Nevertheless, *C. grandis* is so specific that it is unlikely to have an effect on insect species other than the boll weevil.

Life Cycle

Catolaccus grandis is an idiobiont parasitoid, e.g., it arrests host development upon parasitism. Some of its characteristics relating to this trait include:

- **Ectoparasitism.** Ectoparasitoids feed on their host externally by attaching themselves to the integument (outside covering) of the host. These parasitoids are not intimately dependent on host endocrinology (hormonal changes) and behave almost as a predator with the difference that they require only one host to complete development. Endoparasitoids, on the other hand, live inside the host while it continues its development.
- **Host cancelled.** This refers to hosts that live inside plant tissues or create a cavity in the host plant where they live. In general a host is considered to be cancelled if there is a barrier between it and the exterior. Many parasitoids of cancelled hosts, *C. grandis* among them, will not recognize the host if it is presented without a barrier between them. The host's being protected by a barrier is

a requisite because hosts are permanently paralyzed by idiobiont parasitoids making them vulnerable to predators.

- **Long-term host paralysis.** Idiobiont parasitoids paralyze their host by potent venoms or symbiotic viruses (polydnavirus). The purpose of this is to stop host development to allow the immature parasitoids to consume it.

Host stage larger than wasp. Because the host does not develop after being parasitized, it has to be large enough to support the parasitoid larva during its entire development. Only one host is consumed during the development of a wasp.

- **Synovigeny.** This refers to parasitoid females that emerge to adulthood with a limited number of ovarioles and no mature eggs. They have the ability to continuously produce eggs and reabsorb them when a host is not available. On the other hand, females of pro-ovigenic parasitoids emerge with their full complement of eggs already mature and ready to be oviposited.
- **Host-Feeding.** Most adult synovigenic parasitoid females feed on host hemolymph (blood) to obtain additional nutrition for continuous oögenesis (egg production) and egg maturation. Some of these parasitoids require host-feeding to be able to produce the first set of eggs. *C. grandis* females are able to oviposit 4 to 6 eggs without host-feeding, but, they require host-feeding for continuous egg production.
- **Oösorption.** This is the ability of female parasitoids to reabsorb mature eggs when the host is scarce or unavailable. This characteristic goes hand in hand with extended adult longevity. Idiobiont parasitoids live longer, therefore their life cycle does not have to be synchronized with their host life cycle. They are adapted to search and exploit host populations within a large range. They are strong flyers, excellent searchers and are adapted to survive under very low host densities.
- **Ability to chose the sex of progeny according the host size**. Like most pteromalid wasps, *C. grandis* females have the ability to select the gender of their progeny by selectively fertilizing the eggs as they oviposit. They have arrhenotokous reproduction, which mean that all females originate from fertilized eggs while males originate from unfertilized eggs (parthenogenesis). Male eggs are oviposited on hosts that are small in size because males are smaller and require less food to complete development.

In general, idiobiont parasitoids lean closer to being predators than true parasites. They are not linked to host physiology so they must stop host development. They accomplish this by paralyzing the host, but the host must be cancelled to escape predation. They search extensive areas for their host, and adult females live relatively long lives so they are adapted to produce eggs continuously.

Females can locate cotton buds (squares) infested by boll weevils at ranges up to 60 m. Females probe infested squares many times from different angles with their ovipositor. If the square contains a suitable host, the females paralyze the weevil larva and oviposit one single egg inside the cavity next to the paralyzed weevil. The preferred host stages for ovipositing are third instar and early pupa. Females may or may not feed on the host by constructing a feeding tube.

The egg hatches in 18-24 h depending on the temperature. The first instar parasitoid is very active and moves around the cavity until it finds the host, then it attaches externally to feed on it. The larval stage consists of 5 stadia and lasts between 6 to 9 days, the pupal stage lasts between 7 to 9 days. The adults emerge and cut a circular exit hole in the square.

The sex ratio is approximately 3 to 4 females per male. Mating occurs within an hour of emergence and females mate only once. Females have a preovipositional period of 2 to 5 days during which they are not interested in host seeking.

Relative Effectiveness

Augmentative releases of *C. grandis* are very effective against boll weevil infestations. Parasitism rates of 70 to 90% have been reported consistently. The high searching capacity and specificity of this parasitoid makes it ideal for use in inundative releases. Weekly releases of 500 females per acre for 6 consecutive weeks are sufficient to achieve excellent control of the boll weevil, eliminating economical damage by this pest. However, *C. grandis* must be released every year because it does not overwinter in the U.S.

The main obstacle for commercial application of *C. grandis* is the high cost and complexity required for its mass propagation. An artificial diet has been developed to rear *C. grandis*, but a small quantity of host larvae are still required for the conditioning of adult females. Parasitoids produced in vitro have been tested in laboratory and field and they have performed satisfactorily.

Conservation

After inundative release of *C. grandis*, this parasitoid remains in the field as long as there are hosts available for reproduction. The use of nectaried (with extrafloral nectaries) cotton varieties enhances the survival and fecundity of adult parasitoids. The use of mechanical cultivation reduces parasitism rates by *C. grandis* because burying infested squares protects boll weevils from parasitism. Limiting cultivation practices to the beginning of the cotton-growing season reduces its impact on the parasitoids.

Pesticide Susceptibility

As with most parasitoid wasps, *C. grandis* is often more susceptible to insecticides than its host. This parasitoid is highly susceptible to organophosphates such as methyl parathion and malathion. The toxicity to *C. grandis* adults of 10 insecticides including dimethoate, endosulfan, oxamyl, acephate, malathion, azinphos-methyl, cyfluthrin, methyl parathion, spinosad, and fipronil were tested. All of them were highly toxic to the parasitoids although endosulfan showed the least toxicity. Methyl parathion was the most toxic to *C. grandis,* but its low residual effect makes it useful for early applications (prior to releases) against fleahoppers and overwintering weevils. Insecticides that remain toxic in the field for more than a week, such as oxamyl, thiodicarb, and azinphosmethyl were detrimental to the effectiveness of inundative releases.

Cotesia (=Apanteles) glomerata (Hymenoptera: Braconidae)

Cotesia glomerata was introduced to North America in 1883 for the control of the imported cabbageworm on cole crops and has become a major mortality factor of cabbageworm.

Appearance

Cotesia adults are small (about 7 mm), dark wasps and resemble flying ants or tiny flies. They have two pairs of wings, the hindwings being smaller than the forewings, and chewing-lapping mouth parts. The antennae are about 1.5 mm long, and curved (not elbowed) upward. The abdomen of the female narrows to a downward curving extension called the ovipositor with which she lays eggs. The pupae are in an irregular mass of yellow silken cocoons attached to the host larva or to plant leaves.

Habitat (Crops)

Cole crops.

Pests Attacked

The imported cabbageworm and its relative *Pieris brassicae.*

Life Cycle

Adults mate and the females lay eggs, in most cases, immediately after emerging from their cocoons. Eggs are deposited into larvae (preferably first instar) of caterpillars—about 20-60 per larva—soon after mating. A female lays about 150-200 eggs during her life. *Cotesia* larvae emerge after about 15-20 days and spin their cocoons on or near the host which dies when the wasps emerge. The life cycle, from egg to adult, is approximately 22-30 days, depending on the temperature.

Pesticide Susceptibility

In studies *C. glomerata* has been shown to be susceptible to several pesticides commonly used on cole crops.

Relative Effectiveness

In the early season, parasitism by *C. glomerata* is low. However, numbers increase as the year progresses. By midseason, about half of imported cabbageworm larvae are parasitized, and late in the growing season, that percentage has increased to 60-75%. In addition, *C. glomerata* may be an important vector in the transmission of the granulosis virus in imported cabbageworm. It is resistant to attack by many hyperparasitoids.

Conservation

Follow integrated pest management guidelines as presented in this book.

Diadegma insulare (Hymenoptera: Ichneumonidae)

Diadegma insulare is a Lepidoptera parasitoid that occurs naturally in eastern North America to the plains states and in southern California.

Appearance

Diadegma insulare is a small (6 mm long) ichneumonid wasp with reddish-brown legs and abdomen. It pupates inside the cocoon made by the mature diamondback moth larva replacing the host pupal covering with its own cocoon which may have a distinctive white band. Diamondback moth cocoons are white inside (green when the larvae first form the cocoon); *D. insulare* wasps are visible as dark bodies inside the cocoon, before the adult *D. insulare* emerges. Adults can be seen searching in the crop foliage.

Habitat (Crops)

Brassica crops.

Pests attacked

Diamondback moth.

Life cycle

D. insulare overwinters in crop debris as a pupa within the cocoon of its host. Immediately upon emerging, mating takes place, and the female, using her antennae, seeks a host larva (small to medium-sized larvae are selected). When a suitable host has been located, the female flexes her abdomen beneath her thorax and legs and oviposits one egg in each moth larva. She does not sting and paralyze her prey, and larger moth larvae sometimes escape by thrashing about.

After 10 to 15 days, the parasitoid larva emerges from the host cocoon and spins its own cocoon within that of its host. There are 4 to 6 generations per year, corresponding to the number of host generations.

Relative Effectiveness

Diadegma insulare is the most important parasitoid of the diamondback moth in North America and Canada. Native populations of *D. insulare* have parasitized up to 70% of diamondback larvae in field trials in New York and from 50% to almost 90% in Wisconsin.

In one study, diamondback moth larvae placed in agricultural crops (cole crops and a wide variety of other crops including corn, dry beans, alfalfa, tomatoes, and apples), weedy areas, and woodland edges were parasitized in spite of very low (often undetectable) natural numbers of diamondback moth. This suggests that *D. insulare* may have alternate hosts or is a highly effective searcher.

Conservation

Limiting insecticide use and using *Bacillus thuringiensis* (Bt) where possible, allowing wildflowers (especially wild brasiccas) to grow around crop fields, and allowing diamondback moth to colonize wild brassicas and crops such as canola will increase the abundance and effectiveness of *D. insulare* for management of diamondback moth.

D. insulare females require nectar sources. A nectar source can increase *D. insulare* female longevity from 2-5 days to more than 20 days. Numbers of diamondback moth larvae parasitized increases from zero per *D. insulare* female, with a poor nectar source, to more than 150 per female with an optimal nectar source. A number of wildflowers can serve as nectar sources for *D. insulare*, including members of the cabbage family and Queen Anne's lace. Other flowers, including oxeye daisy, perennial sow thistle and common lambs-quarters are poor nectar sources. The best nectar sources were flowers with wide or shallow corollas, where *D. insulare* can easily reach the nectar. However, *D. insulare* also can push between sepals at the base of the flower or chew holes in the sepals, behavior previously reported for bumble bees, but not for ichneumonid wasps. *D. insulare* also can obtain nutrition from aphid honeydew, although it is a poorer food source than flower nectar.

Pesticide susceptibility

D. insulare is very sensitive to insecticides. Adults can be killed by contact with insecticide deposits, and larvae die inside the host, if the host larva is killed by an insecticide, even Bt. However, *D. insulare* may partially avoid insecticide exposure because its pupae are protected inside the diamondback moth cocoon, and the wasps are highly mobile and recolonize treated areas.

Severe diamondback moth outbreaks often involve intensive insecticide use, combined with insecticide resistance in diamondback moth and sensitivity in *D. insulare*. Diamondback moths in the north central U.S. are susceptible to insecticides or are only moderately resistant. However, diamondback moths that are highly resistant to insecticides and also resistant to Bt can be brought into the field on *Brassica* transplants. A grower can quickly get into a downward spiral of increased insecticide use with minimal effect on the diamondback moth larvae followed by reduced parasitism. Thus, the first step in conserving *D. insulare* populations is to carefully inspect transplants for diamondback moth larvae and use insecticides, including Bt, only as needed.

Encarsia formosa Hymenoptera: Aphelinidae

Encarsia formosa Gahan is used worldwide for commercial control of whiteflies in greenhouse crops. Commercial use began in Europe in the 1920s, but by 1945 interest waned due to the development of pesticides. After 1970, use was reinitiated and has expanded from 100 hectares of greenhouse crops to 4800 hectares in 1993.

Comparison of the greenhouse area in various parts of the world with the area employing biological control agents shows that most usage of *E. formosa* occurs in Europe and Russia and that the largest concentrations of greenhouse production in which E. formosa is not extensively used are in North America and Asia, particularly Japan. These are areas where increased use of E. formosa would be possible.

Encarsia formosa was originally described from specimens reared from an unidentified aleyrodid on geranium (*Pelargonium* sp.) in 1924 in a greenhouse in Idaho (USA). There are no synonyms in the literature. Morphological descriptions of all life stages are provided by Speyer. Because of releases into greenhouses worldwide, *E. formosa* has a cosmopolitan distribution and its native range is uncertain.

Appearance

*Encarsia formosa*_females are small (~ 0.6mm in length), have a black head and thorax and yellow abdomen. Males are rare and dark in color.

Habitat

Principal greenhouse crops in which *E. formosa* is used include tomato (*Lycopersicon lycopersicus*) and cucumber (*Cucumis sativus*). The parasitoid is also used, or is being tested, on much smaller areas of eggplant (*Solanum melongena* var. *esculenta*) and gerbera (*Gerbera jamesonii*), poinsettia (*Euphorbia pulcherrima*), marigolds (*Tagetes erecta*), and strawberry (*Fragaria* X *ananassa*). Virtually nothing is known about the ecology of *E. formosa* in outdoor agricultural systems.

Pests Attacked (Host Range)

Encarsia formosa parasitizes at least fifteen species of whiteflies in eight genera. Most work has looked at the ability of *E. formosa* to control greenhouse whitefly *Trialeurodes vaporariorum*, sweetpotato whitefly, *Bemisia tabaci*, and silverleaf whitefly, *Bemisia argentifolii* (= *Bemisia tabaci* strain B). *Encarsia formosa* is hyperparasitized by *Signiphora coquilletti*, *Encarsia pergandiella*, and *Encarsia tricolor*.

Life Cycle

To successfully reproduce in greenhouses, *E. formosa* must locate potential hosts, assess host quality, and use nymphs appropriately for host feeding or parasitism. Following release into the hosts' habitat (i.e., greenhouses), *E. formosa* employs visual and olfactory cues to find infested host plants. When searching new leaves, the parasitoid does not distinguish between upper and lower surfaces and shows no preference for middles or edges of leaves. The rate at which hosts are encountered is dependent on the parasitoids' walking speed, whitefly size, and number of hosts on a leaf. Walking speed is reduced by leaf venation, high trichome densities, excessive honeydew, encounters with nymphs suitable for host feeding and parasitism, decreasing temperature, low barometric pressure, and smaller egg loads.

Encarsia formosa is a solitary endoparasitoid that matures 8-10 eggs per day. Daily egg maturation and oviposition rates decline as wasps age. Adults obtain energy and nutrients by consuming honeydew and hemolymph of hosts that are pierced with the ovipositor, but in which no egg is deposited.

Killing hosts for nutritional purposes is termed host feeding. *Encarsia formosa* will host feed on all pre-imaginal stages of *T. vaporariorum* except the egg, but prefers second instar nymphs and pupae. However, the pupae and all nymphal stages of *B. tabaci* are used equally for host feeding. To host feed, *E. formosa* wounds nymphs or pupae by probing with the ovipositor for up to six minutes and feeds from wounds which wasps may enlarge with their mandibles. This probing followed by feeding kills hosts. Nymphs that have been used for feeding are not used for oviposition, and previously parasitized whiteflies are not used for host feeding (Hoddle et al. 1998).

Encarsia formosa will oviposit in all immature stages of *T. vaporariorum,* except the egg and the mobile first instar, and in all immature stages of *B. tabaci* older than the settled first instar nymph. *Encarsia formosa* prefers to oviposit in third, fourth, and prepupal nymphs of both *T. vaporariorum* and *B. tabaci*. The rate of successful emergence of the parasitoid is highest from these preferred stages. *Encarsia formosa* does not oviposit in up to 50% of suitable hosts in the preferred stages even when these are not parasitized or mutilated from host feeding. Such hosts may be parasitized at a later encounter. Failure to oviposit in such hosts may result from defensive host movements.

Encarsia formosa reared on *T. vaporariorum* can lay five eggs per day (will oviposit a total of 59 eggs before death), host feed on three nymphs per day, and kill on average a total of 95 nymphs over a 12 day life expectancy. Adult females chew a round exit hole on the dorsal surface of fourth instar nymphs before emerging. At 21°C, and with third instar *T. vaporariorum* as hosts, the time from oviposition to adult emergence is 25 days.

Thelytoky (parthenogenetic reproduction in which only females are produced) in *E. formosa* is mediated by *Wolbachia* bacterial infections. Exposure of females to antibiotics or high temperatures (31°C) for two or more generations suppresses microbial activity, allowing females to successfully produce male offspring. Fecundity is reduced once symbionts are eliminated. Males develop as primary endoparasitoids of whiteflies. The mating behavior of *E. formosa* has been described but males are unable to successfully inseminate females.

Relative Effectiveness

Four distinct methods of releasing *E. formosa* into greenhouses for whitefly control have been suggested. Three of these ("pest in first," "dribble," and "banker plants") are inoculative in nature and establish a reproducing parasitoid population, after which releases are discontinued. The fourth approach, in which repeated parasitoid releases are made throughout the cropping season, is used when a reproducing population of parasitoids is not expected to develop, either because the cropping season is too short or the whitefly or host plant are unfavorable. Whitefly mortality results from host feeding or superparasitism.

The "pest in first" method begins with the deliberate introduction of adult whiteflies into greenhouses at a fixed rate (e.g., two whitefly adults per tomato plant). *Encarsia formosa* is later introduced one to three times at a standard rate (e.g., eight parasitized nymphs per tomato plant) at regular intervals which coincide with availability of host stages suitable for parasitism. This method has not been widely adopted due to concern over releasing pests onto the crop.

With the "dribble method", parasitoid introductions begin at planting in anticipation of natural development of a whitefly population. Regular parasitoid releases at a low rate (e.g., one parasitized nymph per plant) continue until parasitized nymphs are found in the crop.

The "banker plant" system utilizes established breeding colonies of whiteflies and parasitoids on earlier grown plants from which wasp and whitefly disperse into the crop. Banker plants are introduced at a fixed rate (e.g., one banker plant per 352 crop plants). Mesh screens can be used to cage banker plants to contain whiteflies while allowing the smaller adults of *E. formosa* to migrate into crop production areas.

Inundative programs require regular releases of high numbers *E. formosa*; establishment and reproduction of the parasitoid in the crop are not expected. This method is applied most frequently to ornamental crops.

The pest in first, dribble, and banker plant techniques have provided successful control of *T. vaporariorum* on cucumber and tomato crops. Success in these cases has been defined in relation to levels of sooty mold (*Cladosporium* sp.) contamination of foliage and fruit. If at harvest sooty mold levels are within commercially acceptable limits, adequate control of *T. vaporariorum* is considered to have been achieved.

In floral crops, the presence of whiteflies at even very low densities (e.g., 0.02 to 0.03 nymphs per cm^2 in poinsettias) is considered damaging and market standards require greater levels of whitefly suppression than are used for vegetable crops (e.g., 7.0 nymphs per cm^2 in tomato). Consequently, use of *E. formosa* has been more extensive in vegetables than in floral crops.

Inundative releases of *E. formosa* have been successful in some instances for control of *T. vaporariorum* on poinsettia. Control of *B. argentifolii* with weekly releases of more than three adult parasitoids per plant per week, has not been accomplished. However, control of this whitefly species has been reported with lower weekly release rates (under two parasitoids per plant), or when *T. vaporariorum* co-occurred in the crop. In one study, as the number of parasitoids released per plant increased, parasitoid efficacy decreased, and *B. argentifolii* survivorship increased.

Conservation

The conservation of natural enemies is probably the most important and readily available biological control practice available to growers. Natural enemies occur in all production systems, from the backyard garden to the commercial field. They are adapted to the local environment and to the target pest, and their conservation is generally simple and cost-effective. With relatively little effort the activity of these natural enemies can be observed. Lacewings, lady beetles, hover fly larvae, and parasitized aphid mummies are almost always present in aphid colonies. Fungus-infected adult flies are often common following periods of high humidity. These natural controls are important and need to be conserved and considered when making pest management decisions. In many instances the importance of natural enemies has not been adequately studied or does not become apparent until insecticide use is stopped or reduced. Often the best we can do is to recognize that these factors are present and minimize negative impacts on them. If an insecticide is needed, every effort should be made to use a selective material in a selective manner.

Pesticide Susceptibility

More than seventy articles have been published that examine interactions between *E. formosa* and one or more pesticides, either in laboratory tests or under conditions of practical use in greenhouses. Standardized methods for determining the effects of pesticides on *E. formosa* have been developed and the effects of more than one hundred different compounds on *E. formosa* have been determined. Selective materials of interest for possible combination with *E. formosa* include insecticidal soap, buprofezin, azadirachtin, abamectin, and resmethrin.

Encarsia inaron (Hymenoptera: Aphelinidae)

Encarsia inaron is a tiny parasitoid wasp that was collected in Italy and Israel and introduced into California in 1989 for the control of the ash whitefly, *Siphoninus phillyreae* (Homoptera: Aleyrodidae). Ash whitefly is native to Europe, the Mediterranean, and northern Africa. In both its native home and California it is commonly associated with woody shrubs and trees.

Appearance

Encarsia inaron is a stingless wasp about 0.5 mm in length and difficult to see without a hand lens. Adults appear as minute flies moving about on leaves with whitefly nymphs. Both the adult male and female have dark colored heads and eyes and clear wings, but can be separated by the yellow abdomen of the female and black abdomen of the male. The black pupal case of the parasitoid causes the whitefly nymph to appear black, differentiating it from healthy whitefly pupae that are yellow in color when their wax is removed. The parasitoid pupa chews a round hole upon eclosion to escape from the body of the whitefly. In contrast, healthy whitefly adults push their way out of their pupal case, leaving a "T" shaped split in the whitefly remains.

Habitat

Woody shrubs and trees.

Pests Attacked

Ash whitefly, *Siphoninus phillyreae.*

Life Cycle

Unlike most species of *Encarsia,* both male and female *E. inaron* develop as primary parasitoids on ash whitefly. It is the most common parasitoid associated with ash whitefly in its native home.

Parasitoids oviposit into third and fourth instar whiteflies, and complete their development inside the whitefly host. The developing larvae become banana shaped and translucent in color, inside late instar whitefly nymphs, but can only be seen through dissection and with the aid of a microscope. Parasitoids at 25°C take approximately three weeks to develop from eggs to adults. At the same temperature, adults live two to three weeks, laying about 159 eggs each.

Relative Effectiveness

Prior to the establishment of *E. inaron* in California, ash whitefly infestations in major urban centers resulted in early season defoliation of street trees such as species of ash and ornamental and flowering pear trees. In some neighborhoods the air was filled with clouds of adult whiteflies. Although some generalist natural enemies fed on these whiteflies, they completely lacked specific parasitoids or predators.

By 1992 *E. inaron* had been colonized throughout most of California. As of 1996, ash whitefly continues to be maintained at very low levels and is difficult to find most of the year. Although we have heard some reports of heavily infested trees, these are isolated events. This single introduction is responsible for controlling ash whitefly infesting urban ornamental trees throughout the state and represents classical biological control at its best. Prior to colonization of *E. inaron,* ash whitefly affected millions of homeowners, and received much media coverage. Today, ash whitefly is difficult to find.

Conservation

Encarsia inaron has dispersed over wide areas of California, and apparently other parts of the country, wherever ash whitefly has moved. Once established, the parasitoid regulates densities of ash whitefly to very low numbers thus additional control of this pest is in almost all cases unnecessary. The best way to conserve the parasitoid is to avoid pesticide usage in or near trees where *E. inaron* is present. In the rare case in which ash whitefly may reach high numbers, spraying the foliage repeatedly with water or a soap spray is recommended. The parasitoids overwinter in both the duff of trees and on some evergreen shrubs infested by ash whitefly during winter months, such as citrus, pyracantha and toyon. Maintaining fallen leaves at the base of deciduous trees susceptible to ash whitefly (ash and ornamental pear trees) or planting the above shrubs nearby may provide increased continuity in the parasitoid's presence.

Eretmocerus eremicus (=*Eretmocerus* sp. nr. *californicus*, Arizona strain) (Hymenoptera: Aphelinidae)

Eretmocerus eremicus is a tiny parasitic wasp (~1 mm in length) that is indigenous to the southern desert areas of California and Arizona and is an important parasitoid of whiteflies in these areas.

Appearance

Female *E. eremicus* are pale lemon yellow with green eyes and clubbed antennae. The name *Eretmocerus* is derived from Latin, meaning "oar-like," and refers to the shape of the female antennae. Male wasps have longer, elbowed antennae, and are yellowish brown in color. Whitefly pupae that have been parasitized by *E. eremicus* appear beige in color whereas healthy whitefly pupae are white or pale yellow. This aphelinid parasitoid does not deposit fecal-like material within the host before emergence as do some other parasitoids of whiteflies. *E. eremicus* can only emerge through the upper surface of the host. This is accomplished by chewing a circular exit hole.

Habitat (Crops)

E. eremicus have been released inundatively in the Imperial Valley of California, to control *Bemisia argentifolii* on cotton. Inundative releases have also been made to suppress whiteflies infesting ornamental and vegetable crops in greenhouses.

Pests Attacked

E. eremicus attacks whiteflies (Homoptera: Aleyrodidae) including greenhouse whitefly (*Trialeurodes vaporarium*), sweetpotato whitefly (*B. tabaci*), silverleaf whitefly (*Bemisia argentifolii*), and bandedwinged whitefly (*T. abutlonea*).

Life Cycle

Both male and female *E. eremicus* develop as primary parasitoids of whiteflies. In commercial mass rearing with greenhouse whitefly as the host, this wasp has a sex ratio of 1:1 (female:male), however in the field the sex ratio may favor females. *E. eremicus* females locate whitefly hosts, in part, by olfaction. Females lay their eggs between the whitefly nymph and the leaf surface. Females will oviposit under all immature whitefly stages, except eggs, but second instars may be preferred.

The *E. eremicus* egg hatches about 4 days after being laid (the exact time is dependent on temperature) and the wasp larva develops as an external-internal parasitoid. The newly emerged larva attaches its hook-like mouthparts to the underside of the whitefly nymph and chews a small hole into the whitefly. After 3-4 days of chewing the parasitoid larva enters the host where it remains dormant until the whitefly pupates. Once the whitefly pupal stage is reached, the wasp larva releases digestive enzymes, and begins ingesting the semi-liquid body parts of the pupa.

The wasp larva passes through three instars, requiring about 12 days to complete development. The adult wasp chews its way out of the whitefly cadaver. Adult female wasps can live for 6-12 days at 27°C. Longevity depends on temperature and availability of food, such as honeydew produced by whiteflies. Females lay 3-5 eggs per day. Female wasps also kill whitefly nymphs by repeatedly probing with their ovipositors and feeding on the haemolymph (blood) that exudes from the wound.

Relative Effectiveness

E. eremicus is an aggressive searcher, covering around 1.3 mm/s on some plants. Females will inspect all whitefly stages with the same relative frequency as their encounter rate. Around 74% of nymphs that are probed with the ovipositor are parasitized. Optimum temperatures which will facilitate rapid development and egg production occur at 25-29°C. In the greenhouse environment, the temperature should be manipulated to ensure that *E. eremicus* will develop more quickly than its host. On greenhouse grown poinsettia in the northeastern US, a release rate of 3 female *E. eremicus*/plant/week achieves very high levels of mortality (> 98%) of silverleaf whitefly in the first 6-8 weeks of the crop.

Pesticide Susceptibility

E. eremicus is assumed to be susceptible to pesticide residues and fumigants. Releases of *E. eremicus* will be most effective in crops that have not been sprayed for 10-14 days. Employ the *Encarsia formosa* recommendations for *E. eremicus* from the Koppert side effect list of pesticides for beneficial organisms (available from Koppert Biological Systems).

Eriborus terebrans (Hymenoptera: Ichneumonidae)

The ichneumonid wasp, *Eriborus terebrans,* was introduced into the United States as part of a classical biological control project to control the European corn borer. Approximately 140,000 wasps collected from both Asia and Europe were released from 1927 through 1940 in 13 states from Vermont to Virginia and as far west as Indiana and Michigan.

E. terebrans is one of 6 of these introduced species to become established and is one of 3 considered to be important in controlling European corn borer. (*Macrocentrus grandii,* a braconid wasp, and *Lydella thompsoni,* a tachinid fly, are the others.) It is currently the most frequently recovered parasitoid of European corn borer in the north central region of the United States.

Appearance

A medium-sized wasp, female *E. terebrans* are approximately 6-10 mm in body length. Males are slightly smaller. Females have a curving ovipositor (about 3 mm) protruding beyond the abdomen, and both sexes have long antennae (5-7 mm) and are black with reddish-brown legs. The smooth abdomen is slightly more shiny than the head and thorax.

Habitat

Corn fields and surrounding habitats.

Pests Attacked

Ostriana nubilalis, the European corn borer.

Life Cycle

E. terebrans overwinters as a larva inside overwintering corn borers and emergence occurs from 22 to 37 days after resumption of development in the spring. First generation emergence of wasps is in synchrony with the first larval generation of European corn borers. Females mate soon (sometimes within an hour) after emergence and can lay eggs within one day. They prefer to lay their eggs in 2nd-4th stage corn borers.

Adult wasps may live 7-10 days under ideal conditions (75-80°F, and with access to water and sugar). However, lifespan is greatly reduced and is only 3-4 days when temperatures are above 90°F and there are no sugar sources such as flower nectar or aphid honeydew available.

Females are highly attracted to chemicals in corn borer frass and webbing, and to corn borer cuticle, larvae, oral secretions, and feces. When exposed to corn borers, they exhibit ovipositing behavior. *E. terebrans* is also attracted to corn plants, the host of the European corn borer.

Relative Effectiveness

One of the most widely distributed parasitoids of the corn borer North America, *E. terebrans* parasitized an average 4.9% and 18.7% of first generation corn borers, and 10.2% and 9.1% of second generation corn borers in Michigan in 1989 and 1990, respectively. Although the activity peak of first generation *E. terebrans* adults is well synchronized with the population peak of first generation corn borers, second generation wasp emergence peaks about 4 weeks earlier than the 2nd generation corn borer population.

The highest level of parasitism by *E. terebrans* observed in individual fields in the Michigan study was 37.4%, occurring during the first generation in 1990. In 1938 near Boston, reported parasitism was 55.8%, but long term averages of parasitism in the north central region of the U.S. are 2.4%-7.8%. It should be noted, however, that these figures are from second generation parasitism.

Previous research has suggested that first generation corn borers were more likely to be parasitized by *E. terebrans* than second generation larvae, but this may vary from year to year. Landis and Haas reported that in most fields during 1989 and 1990, *E. terebrans* parasitized more first generation corn borers on field edges where fields bordered wooded areas. This relationship was only seen in fields bordering wooded areas, and it was not seen at all with second generation corn borers regardless of bordering vegetation.

Additional studies showed that adult wasps die rapidly when temperatures exceed 90°F and that wasps require some type of sugar on a daily basis or they will die. During the first corn borer generation, prior to corn canopy closure, maximum temperatures in corn fields may often exceed 90°F, and sources of nectar or honeydew may be scarce in the middle of corn fields. Wasps are able to survive better in wooded field borders, where there is more shade and often flowering plants or sources of aphid honeydew. First generation European corn borer larvae near wooded edges were parasitized at two to three times the rate of those in field interiors. Adults are not known to feed on hosts.

If corn is planted in rotation with other crops, *E. terebrans* adults must disperse from their overwintering sites to the current corn fields before they can become a factor in corn borer control.

Conservation

Wooded areas adjacent to corn fields where *E. terebrans* can find nectar and/or aphid honeydew and a refuge from the heat will help adult wasps endure high temperatures.

The conservation of natural enemies is probably the most important and readily available biological control practice available to growers. Natural enemies occur in all production systems, from the backyard garden to the commercial field. They are adapted to the local environment and to the target pest, and their conservation is generally simple and cost-effective. With relatively little effort the activity of these natural enemies can be observed. Lacewings, lady beetles, hover fly larvae, and parasitized aphid mummies are almost always present in aphid colonies. Fungus-infected adult flies are often common following periods of high humidity. These natural controls are important and need to be conserved and considered when making pest management decisions. In many instances the importance of natural enemies has not been adequately studied or does not become apparent until insecticide use is stopped or reduced. Often the best we can do is to recognize that these factors are present and minimize negative impacts on them. If an insecticide is needed, every effort should be made to use a selective material in a selective manner.

Pesticide Susceptibility

Unknown.

Lysiphlebus testaceipes Hymenoptera: Aphidiinae

Aphidiid wasps attack only aphids. The conspicuous sign of aphidiid activity is the presence of aphid "mummies" - swollen, dead aphids that have been tanned and hardened to form a protective case for the developing wasp pupa.

Appearance

L. testaceipes is a tiny (<3mm) black wasp and is not commonly seen. However, the distinctive aphid mummies which remain on leaves after the parasitoid has killed the greenbug can easily be detected. The mummy consists of the outer skin of the greenbug which becomes modified into a tough protective shell after the developing wasp kills the greenbug by its internal feeding. Greenbugs parasitized by *L. testaceipes* are beige or tan in color and are round and swollen compared to healthy greenbugs.

Habitat (Crops)

Wheat and grain sorghum.

Life Cycle

L. testaceipes overwinters as a grub or pupa inside a parasitized aphid. The newly emerged wasp mates, and then begins to search for new aphids to attack. The female wasp inserts an egg into the greenbug and in about 2 days a tiny wasp grub hatches and feeds internally on the living aphid. The wasp grub completes feedings in about 6-8 days, resulting in the death of the aphid. Movement of the wasp grub inside the aphid expands the aphid, giving it a swollen appearance. The larva cuts a hole in the bottom of the aphid, attaches the aphid to a leaf with silk and a glue, and the dead greenbug changes color from green to a brown "mummy". Then the wasp grub molts to the pupal stage, and after 4-5 days a wasp emerges by cutting a circular hole in the top of the mummy.

At 70°F, development from egg to adult takes about 14 days. Wasps disperse by flying, or by being carried inside winged aphids which may undergo long migration flights.

Relative Effectiveness

Wasp parasites contribute to greenbug suppression in two ways. There is direct mortality caused by the wasp parasitism, but, in addition, parasitized aphids have reduced reproductive rates. Parasitized greenbugs stop reproducing within 1-5 days, while healthy greenbugs give birth to 3-4 live greenbugs a day for 25-30 days. Thus, the activity of these wasps can greatly reduce the rate of greenbug increase.

Hyperparasite attack may reduce the effectiveness of *L. testaceipes* if the hyperparasite is abundant.

Parasitoid activity in the field can be monitored by looking for greenbug mummies on crop leaves. As a general rule, a greenbug infestation usually declines rapidly after 20% of the greenbugs are mummies; at this point most of the living greenbugs have been parasitized but have not yet turned into mummies. Normally mummies appear 8-10 days after wasps lay their eggs in the greenbug.

Temperature is an important factor influencing the efficacy of wasps as biological controls of greenbugs. Wasps develop most rapidly when temperatures are above 65°F, and adults are not active if temperatures are below 56°F. However, greenbugs are much more tolerant of cool temperatures and continue to reproduce until temperatures drop to 40°F. Thus wasps may not be effective in controlling greenbugs in wheat in the fall and spring due to cool weather.

Pesticide Susceptibility

Pesticide use in wheat or grain sorghum may decrease activity of these parasitic wasps. Insecticides applied as sprays will kill adult wasps as well as immature wasps developing inside greenbugs killed by insecticides.

Research in Texas has shown the methyl parathion and chlorpyrifos are more toxic to adult wasps and to immature wasps inside greenbugs than systemic insecticides such as dimethoate or disulfoton, especially at lower rates. However, the shorter residual activity of methyl parathion allows parasites to recolonize a field sooner after treatment. A fungicide (triadimefon; Bayleton) used to control leaf rust in wheat is also very toxic to adult wasps.

Macrocentrus cingulum (=grandii) Brischke (Hymenoptera: Braconidae)

In 1926, *Macrocentrus cingulum* was introduced to the U.S. from France and Korea along with a range of other exotic natural enemies to control European corn borer (*Ostrinia nubilalis*), a moth whose caterpillars damage corn plants by boring into their stems. *Macrocentrus cingulum* attacks caterpillars of the European corn borer, exerting low to moderate levels of parasitism in the field. *Macrocentrus cingulum* may also be referred to by its synonymous (i.e. equivalent) names, *M. grandii* Goidanich, *M. abdominalis* Fab. or *M. gifuensis* Ashmead.

Appearance

Adults are usually 4 mm in length, and range in color from black to brick red. The head is usually dark brown to black. Adults have an elongated abdomen and a long egg-laying structure (ovipositor) protruding from the rear. The top of the abdomen usually has a pale colored band.

European corn borer

Habitat

Macrocentrus cingulum has been recorded from Europe, Japan, Korea and China, and was introduced to North America. It is the most abundant parasitoid of the European corn borer in the Midwestern and Northeastern U.S. In the lab, individuals of *M. cingulum* are attracted to odors released by corn, and to a lesser extent, potato and snap beans. This behavioral flexibility could allow it to locate its host, the generalist European corn borer, as it feeds on different plants. However, in nature *M. cingulum* has only been found attacking the European corn borer in cornfields.

Pests attacked

The main host for *M. cingulum* in North America is the European corn borer (*Ostrinia nubilalis*). However, it also parasitizes the Asian corn borer, *Ostrinia furnacalis*, in its native range in Eurasia.

Life cycle

Females of *M. cingulum* lay eggs on 2nd to 4th instars (immature stages) of the European corn borer. Opportunities to lay eggs are often scarce, because the corn borer uses the cavity it excavates in the corn stalk as a refuge. This species is polyembryonic, meaning that many individuals can develop from one egg and are all clones (genetically identical) to each other. On average, 25-26 wasps emerge from one corn borer host.

This species has a total of four larval stages. The first three of these feed on the European corn borer host from within. The fourth immature stage emerges from the host when the corn borer becomes a pupa, and finishes feeding on the host from the outside. Each individual of *M. cingulum* spins its own cocoon near the body of the host; together these form a mass of spun cocoons. After remaining in the pupa for some period of days, the individuals all emerge around the same time. In the lab, adult females live around 21 days while males live 14 days, but survival in the field is likely to be shorter. There are two generations of *M. cingulum* per year.

Relative Effectiveness

The effectiveness of *M. cingulum* ranges from low to moderate, depending on location, infection by pathogens and other factors. In Pennsylvania, around 17% of corn borers were parasitized by this wasp, while in Illinois, the rate of parasitism ranged from 4%-23%. *Macrocentrus cingulum* has been shown to transmit the protozoan *Nosema pyrausta* between individuals of its host. This protozoan causes disease in corn borer. Consequently, *M. cingulum* could provide a service by increasing the spread of this pathogen. From 17-75% of *M. cingulum* were infected with this protozoan, suggesting a high potential for *M. cingulum* to transmit this disease in corn borer populations. However, this parasite can also decrease the lifespan of the adult wasp and depress reproduction. Thus, infection by *N. pyrausta* may be one of the causes of low or decreasing effectiveness of *M. cingulum* in controlling European corn borer populations.

One of the attempts to increase the effectiveness of *M. cingulum* focused on managing the habitat surrounding fields. This wasp feeds primarily on nectar from native flowers, such as milkweed and ironweed. It is possible that planting these additional types of floral resources along field edges may increase the survival of *M. cingulum* and increase its control of European corn borer.

Pesticide Susceptibility

This parasitoid is susceptible to microbial infections, including *N. pyrausta*, for reasons elaborated above. Transgenic Bt (insect resistant) corn and the insecticide permethrin were both shown to decrease the abundance of *M. cingulum*, likely as a result of decreased host availability. Pesticide susceptibility has not been evaluated.

Metaphycus alberti (Hymenoptera: Encyrtidae)

Metaphycus alberti (Howard) was originally brought to California from Australia in 1898 by Albert Koebele, whose earlier entomological investigations of that continent led to the successful biological control of the cottony cushion scale.

Appearance

M. alberti is a member of the *asterolecanii* (= *hederaceus*) group of *Metaphycus*, with two segmented maxillary and labial palpi (Annecke, 1981). *M. alberti* is distinguished from other members of the *asterolecanii* group (for instance, *M. helvolus*) by ocelli that form an acute triangle, and mid-tibia that are significantly longer than their ovipositor in females.

M. alberti is very small - 1.5-2 mm long, and males are noticeably darker than females.

Habitat

Currently, *M. alberti* is a rather rare natural enemy of *Coccus hesperidium* (soft brown scale). Indoors, in locations where it has become established through innoculative biological control programs, it may be found in the vicinity of plants attacked by its host in greenhouses, conservatories and atriums . Plant species preferred by *C. hesperidium* are evergreen, tropical and semitropical species of plants, although it attacks nearly every type of plant except grasses.

Pests Attacked (Host Range)

This species has not been reported in the literature from other hosts, and therefore appears to be specific to *C. hesperidum*. It is known to attack *C. hesperidum* between the crawler and the adult stages, preferring young scale from 1 to 1.5 mm long, but successfully attacking much larger hosts as long as they are not reproducing.

Koebele's original material was reared from *C. hesperidum* collected in the Sydney area.

The parasite was apparently colonized in Riverside, California, around the turn of the century. It was subsequently recovered from *C. hesperidum* by Timberlake during 1911 and 1912 . Despite all of the subsequent sampling of *C. hesperidum* in southern California, *M. alberti* has not been reported from North America since Compere reared it from *C. hesperidum* in 1922. At present, *M. alberti* is known only from California, South Africa, and Australia.

Life Cycle

Biological studies of *M. alberti* have never been conducted. Laboratory rearing experience with this species suggests that it has a life cycle similar to *M. luteolus* or *M. flavus* - two species with which it also shares many morphological and behavioral features.

M. alberti is biparental, develops gregariously [more than one egg in the same host], and has a short developmental period, emerging in as little as 12 days after oviposition. It is easily reared on midsize, immature *C. hesperidum* utilizing methods developed by Bartlett.

Relative Effectiveness

Metaphycus alberti is an active searcher, has a short, uncomplicated life cycle, and is consistently able to overcome egg encapsulation and develop successfully in Texas populations of *C. hesperidum* (Stauffer, 1996). It has been used several times to control entrenched populations of *C. hesperidum* in interior plantscapes. In each case, it provided excellent, long-lasting control.

Pesticide Susceptibility

This parasite is easily killed by pesticides. No pesticide tolerant strains are known.

Muscidifurax raptor (Hymenoptera: Pteromalidae)

An important control agent of filth flies, the parasitoid *Muscidifurax raptor* is thought to have been introduced to the western hemisphere along with the house fly and stable fly, which it parasitizes. It belongs to a large family of about 340 species and is the most effective parasitoid of flies on farms in the northeastern United States.

Appearance

M. raptor is a minute, black wasp, about 1-2 mm long.

Habitat

Dairy farms and poultry ranches, as well as suburban and urban localities where flies exist. *M. raptor* does well both in- and outdoors, and is best suited to controlling fly populations in the northeastern United States.

Pests Attacked

The house fly, the stable fly, and other fly species.

Life Cycle

The adult parasitoid stings the fly pupa, killing the pupa and then lays an egg in the pupal case. When the egg hatches, the larva feeds on the dead fly pupa. In 19-21 days, the *M. raptor* adult emerges from the fly pupal case and begins its search for fly pupae on which to feed and deposit eggs.

Adults feed on fly pupae body fluids. The female may sting the pupa, killing it, without depositing an egg. A droplet of fluid appears at the sting site, and the female feeds on the fluid and may share with a nearby male.

Relative Effectiveness

Flies live longer, develop into adults in half the time, and lay more eggs than *M. raptor* and are thus more numerous than the parasitoid. For control of flies, it is helpful to augment parasitoid populations in early summer. The optimal time to begin parasitoid releases is mid- to late May, and they should continue on a weekly basis until mid-August. In research trials, costs were more than offset by reductions in insecticide expenses.

In one study, adult parasitoids successfully developed in 70% of parasitized pupae, but even when adults failed to emerge, the fly pupae were killed. In addition, due to the adult feeding activity, more flies are killed than are actually parasitized.

Pesticide Susceptibility

M. raptor, having a longer life cycle, has not developed the resistance

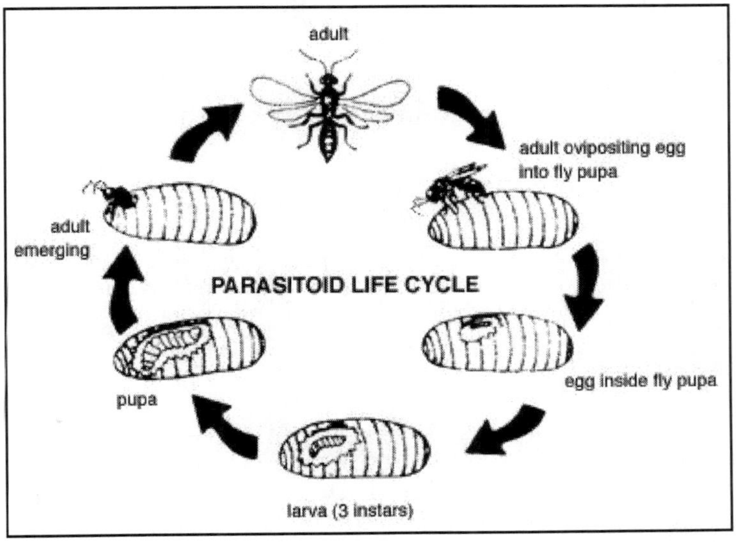

Parasitoid Life Cycle

to commonly used pesticides that flies have developed. In many cases a pesticide will scarcely affect the fly population but will destroy its parasitoids. Research has shown that, in general, *M. raptor* is more susceptible than flies to insecticides. If insecticides must be used, space sprays and baits are most compatible with fly parasitoids.

Conservation

The careful, proper use of insecticides (if necessary) and weekly releases of *M. raptor* along with proper manure management, should help establish a favorable balance of this parasitoid in relation to filth flies.

Nealiolus curculionis (Hymenoptera: Braconidae)

Sunflower, an important oilseed crop in the Northern Plains states, is one of only a few cultivated crops that are native to North America. Insects associated with the sunflower genus *Helianthus* - plant feeders, pollinators, and natural enemies - have evolved with the plant for centuries and many have moved from wild species to the cultivated crop. Only a small number of the hundreds of insects recorded from sunflower have become economic pests; indigenous natural enemies have been a significant factor in preventing many insects from becoming pests.

Appearance

N. curculionis is a small (~3 mm long), dark wasp that is seen on all parts of the sunflower plant, but most often probing the stems. Parasitized sunflower stem weevils are not visibly distinguishable from healthy larvae.

Habitat (Crops)

Sunflowers, both wild and cultivated, and other plants where its several hosts are found.

Pests Attacked

Sunflower stem weevils (*Cylindrocopturus adspersus*) and at least 10 other species in the Curculionidae, including the red sunflower seed weevil (*Smicronyx fulvus*), the boll weevil (*Anthonomus grandis grandis*), and the plum curculio (*Conotrachelus nenuphar*).

Life Cycle

N. curculionis is well-synchronized with its sunflower stem weevil host, suggesting that it has a high degree of adaptation with its host's physiology. *N. curculionis* overwinters as a late first or early second instar within the mature diapausing sunflower stem weevil larva in a chamber constructed in the lower stem or root crown of the sunflower plant. *N. curculionis* does not complete development in one season, and its host continues to function after parasitization. After diapause, the parasitoid larva develops for approximately 21 days, then exits its host, continues growing and feeding externally on the stem weevil (sometimes consuming the entire weevil larva), and pupates. From the end of diapause to adult emergence is usually about 31 days. Only a single adult normally emerges from a host larva, the males approximately 4 days earlier than females.

N. curculionis adults were observed from late June to late August during 1980 to 1985 in the northern plains. Early instar sunflower stem weevils feed and tunnel in the conductive stem tissue which is located under the epidermal surface of the plant stem and it is there, within reach of *N. curculionis'* ovipositor, that parasitoid oviposition takes place. Adults in the study were provided honey and moisture and survived about 9 days at 26°C. (In the field, nectar is plentiful on sunflower, so nutrients are not likely a limiting resource.)

In two of the observation years (1981 and 1984), there were two peaks in the number of *N. curculionis* adults in field plots. This could be due to late season emergence of adults of this generalist parasitoid species from other overwintering curculionid species.

Relative Effectiveness

In a North Dakota study of sunflower stem weevils from 1980 to 1991, parasitism rates by *N. curculionis* ranged from about 5% to almost 32% (average = 27%), and comprised about 96% of the parasitoids attacking the weevils. *N. curculionis*is also the predominant parasitoid of sunflower stem weevils in Minnesota and South Dakota.

Study results did not show a positive correlation between parasitism rates and host density, indicating that the female parasitoid is able to locate and attack hosts under varying host population densities. The degree of mortality from the host's immune response also varies from year to year due to variation in the intensity of the host's defensive reaction over time.

The parasitoid appears to be a consistent mortality factor in the population dynamics of the sunflower stem weevil in cultivated sunflower, though adult population levels are often low.

Conservation

The conservation of natural enemies is probably the most important and readily available biological control practice available to growers. Natural enemies occur in all production systems, from the backyard garden to the commercial field. They are adapted to the local environment and to the target pest, and their conservation is generally simple and cost-effective. With

relatively little effort the activity of these natural enemies can be observed. Lacewings, lady beetles, hover fly larvae, and parasitized aphid mummies are almost always present in aphid colonies. Fungus-infected adult flies are often common following periods of high humidity. These natural controls are important and need to be conserved and considered when making pest management decisions. In many instances the importance of natural enemies has not been adequately studied or does not become apparent until insecticide use is stopped or reduced. Often the best we can do is to recognize that these factors are present and minimize negative impacts on them. If an insecticide is needed, every effort should be made to use a selective material in a selective manner.

Pesticide Susceptibility

Materials that were the most effective in controlling stem weevil larvae in one study were also most harmful to the parasitoids. The majority of these compounds were carbamates, which tend to be more persistent that organophosphates. This may be one of the reasons for their destructive effect on the natural enemies. Carbofuran was the most toxic insecticide tested to both the parasitoid and its host.

The non-target effect of these materials on natural enemies must be weighed against the benefits of their activity on stem weevils in the development of any pest management program.

Peristenus digoneutis Loan (Hymenoptera: Braconidae)

Peristenus digoneutis was discovered in northern Europe by staff of the USDA European Biological Control Laboratory (Montpellier, France). There it was an effective parasitoid of *Lygus rugulipennis Poppius*. Because the latter insect is an ecological equivalent of *Lygus lineolaris* (Palisot) and *L. hesperus* Knight, two closely-related North American pests of many crops, and these tarnished plant bugs lack effective native parasites, *P. digoneutis* was released against *L. lineolaris* in the northeastern US by the USDA Beneficial Insects Laboratory, Newark, DE, under W. H. Day.

After considerable efforts over several years, *P. digoneutis* was established in northwestern New Jersey alfalfa in 1984. By 1989, it had spread to southeastern New York. It began to move up the Hudson Valley by 1993 and reached the Canadian border (Clinton County) by 1996. It has been found in ten northeastern states (CT, MA, NH, NJ, NY, PA, VT, RI, DE, OH), in three Canadian Provinces, and is still spreading. However, it is not abundant south of latitude 41 ?N (New York City), so another parasitoid species that is adapted to a warmer climate is needed there.

Appearance

P. digoneutis is a small, dark brown, 3 mm (1/8") long wasp. (See photo to right).

Habitat

Alfalfa fields, apple orchards, and strawberry fields, at present. It is hoped that further research will show that other crops attacked by tarnished plant bug provide a habitat for *P. digoneutis.*

Pests Attacked (Host Range)

P. digoneutis has reduced tarnished plant bug numbers in New Jersey alfalfa by 75%, and researchers in New York and New Jersey learned that it will fly into apples and strawberries and parasitize *Lygus* pests there. Other crops damaged by tarnished plant bug need to be studied, such as peaches, raspberries, lima and dry beans, and several seed crops.

Life Cycle

P. digoneutis overwinters as a pupa in its cocoon, in the soil, and the adult wasp emerges in June. Females inject a single egg into each tarnished plant bug nymph; it hatches in 5-7 days. The larva feeds inside the young tarnished plant bug, maturing in 7-10 days. The parasitoid then leaves the dying host and drops to the ground, where it constructs a cocoon just under the soil surface. There are two full generations and apparently a partial third per year, each synchronized with the host's major generations. An increasing proportion of each generation diapause (delay emergence from their cocoons until the following June) to carry the species through possible periods of host scarcity and the winter.

Relative Effectiveness

P. digoneutis has reduced tarnished plant bug numbers in New Jersey alfalfa by 75%, and cooperators in New York and New Hampshire are doing research to learn if it will fly into strawberries and parasitize *Lygus* pests there or if alfalfa must be nearby. If the parasitoid is found to disperse, other crops damaged by tarnished plant bug will be studied, such as peaches, apples, raspberries, lima and dry beans, and several seed crops.

In northwestern New Jersey, parasitism rates reached 55%, and tarnished plant bug numbers in alfalfa promptly declined. Now that pest numbers have decreased by 75%, parasitism declined to about 30% (see graph). This successful biological control research should benefit agriculture in two ways – directly by reducing the slight feeding losses to alfalfa, and indirectly by reducing the number of tarnished plant bug adults which fly into valuable fruit and vegetable crops and damage them, especially when nearby alfalfa is harvested. If *P. digoneutis* is found to remain in perennial fruit crops like peaches and apples, even greater biological control of tarnished plant bug should be realized.

Conservation

P. digoneutis persists well in alfalfa fields, despite 3-4 harvests per year. If it is found to be effective in other crops, conservation methods will be developed, if necessary.

The conservation of natural enemies is probably the most important and readily available biological control practice available to growers. Natural enemies occur in all production systems, from the backyard garden to the commercial field. They are adapted to the local environment and to the target pest, and their conservation is generally simple and cost-effective. With relatively little effort the activity of these natural enemies can be observed. Lacewings, lady beetles, hover fly larvae, and parasitized aphid mummies are almost always present in aphid colonies. Fungus-infected adult flies are often common following periods of high humidity. These natural controls are important and need to be conserved and considered when making pest management decisions. In many instances the importance of natural enemies has not been adequately studied or does not become apparent until insecticide use is stopped or reduced. Often the best we can do is to recognize that these factors are present and minimize negative impacts on them. If an insecticide is needed, every effort should be made to use a selective material in a selective manner.

Pesticide Susceptibility

P. digoneutis is seldom exposed to insecticides in alfalfa because all major pests of this crop in the northeastern US but one (potato leafhopper) are already under biological controls developed by the Newark laboratory. It is likely that both *Lygus* and its parasite numbers are low in crops that are regularly treated with insecticides.

Pholetesor ornigis (Hymenoptera: Braconidae)

Pholetesor ornigis is a small braconid wasp native to North America.

Appearance

P. ornigis can be seen and monitored by carefully opening leafminer mines: use a fine-pointed forceps or pin and a 10-power hand lens. Carefully tear back the lower covering of a late-stage spotted tentiform leafminer mine. The leafminer larva should be relatively easy to see. Parasitized leafminer larvae remain active until the fifth instar or prepupal stage. A leafminer killed by *Pholetesor ornigis* will appear as a dead and shrunken larva with a football-shaped, tightly woven silken parasitoid cocoon nearby. The cocoon is pale in color with a dark band around the middle.

Habitat

Northeastern United States and eastern Canada.

Pests Attacked (Host Range)

Spotted tentiform leafminer (*Phyllonorycter blancardella*), apple blotch leafminer (*P. crataegella*), and other gracillarids.

Life Cycle

P. ornigis overwinters as a prepupa within a cocoon in the leaf mine of its host. This occurs concurrent with leaf fall in southeastern Canada. First generation flight of *Pholetesor* occurs during the petal fall period and for several days after. There are 3-4 generations per year and four to five distinct adult flight periods.

Adult females oviposit primarily in sap feeding, 3rd-5th instar larvae of leafminers. When 1st instars are parasitized, *P. ornigis* development time is about 5 days longer than on other instars. This suggests that a delay in parasitoid development takes place until the host reaches the second or later instar. The same delay is reported in other braconids, as well, and allows parasitoid oviposition in hosts that are too small to support development but are also too weak, immunologically and physically, to reject it.

Total development time from egg to adult is about 23 days at 23°C.

Relative Effectiveness

In the laboratory with high leafminer densities, *P. ornigis* females parasitized an average of 370 leafminer larvae, and they are highly fecund. In southern Canada (Guelph), the parasitism rate of spotted tentiform leafminer by *P. ornigis* was found to be 10%.

Because of the closely synchronized life cycle of *P. ornigis* with that of spotted tentiform leafminers, it is a very efficient parasitoid during all three leafminer generations. *P. ornigis* females live two times as long and produce three times as many offspring as its host, the spotted tentiform leafminer.

In Wisconsin studies, *P. ornigis* appeared to contribute less parasitism during the second leafminer generation than the other major parasitoid of spotted tentiform leafminer, *Sympiesis marylandensis*. However, it was found that *Sympiesis* will parasitize leafminer larvae that have already been internally parasitized by *Pholetesor*. Therefore, it is likely that *Pholetesor* would have an even greater impact on the second leafminer generation in the absence of *Sympiesis*.

Conservation

Successful biological control of spotted tentiform leafminer requires conservation of its natural enemies, especially the parasitoids. *Pholetesor* adults are flying during the young sap-feeder stages of the leafminer. Flight periods other than the one occurring during petal fall can be determined by using yellow sticky cards or by monitoring when the leafminer generation is in the sap feeder stage. Spraying should be avoided during all flight periods, if possible.

Although the petal fall insecticide has long been considered critical for managing early season apple pests, it is not always necessary. If leafrollers are a major concern, they can frequently be managed with conventional insecticides before bloom, or with microbial insecticides based on *Bacillus thuringiensis* before bloom or at petal fall.

Codling moth activity often doesn't peak until after petal fall; this can be monitored with pheromone traps. Plum curculio is the remaining major apple pest that can cause damage at and after petal fall. By careful visual monitoring, you can determine if it is necessary to control plum curculio.

Many growers in Wisconsin found that it was not necessary to use the petal fall insecticide every year. By eliminating this spray when possible, not only can money be saved, but *Pholetesor* can be conserved and biological control improved.

Apple growers continue to adopt Integrated Pest Management practices, including careful monitoring of pests and natural enemies and the use of alternative pest control practices. We are seeing a reduction in the use of broad spectrum insecticides that are harmful to beneficials such as *Pholetesor* and an overall decrease in secondary pest problems such as leafminers, aphids, and spider mites. With a little effort and an understanding of the natural enemies and their requirements, conservation biological control of spotted tentiform leafminer can be a reality.

Adults require carbohydrate food such as honeydew from aphids and coccids, tree sap ooze, or floral nectar.

The conservation of natural enemies is probably the most important and readily available biological control practice available to growers. Natural enemies occur in all production systems, from the backyard garden to the commercial field. They are adapted to the local environment and to the target pest, and their conservation is generally simple and cost-effective. With relatively little effort the activity of these natural enemies can be observed. Lacewings, lady beetles, hover fly larvae, and parasitized aphid mummies are almost always present in aphid colonies. Fungus-infected adult flies are often common following periods of high humidity. These natural controls are important and need to be conserved and considered when making pest management decisions. In many instances the importance of natural enemies has not been adequately studied or does not become apparent until insecticide use is stopped or reduced. Often the best we can do is to recognize that these factors are present and minimize negative impacts on them. If an insecticide is needed, every effort should be made to use a selective material in a selective manner.

Pesticide Susceptibility

In laboratory studies, the adult wasp stage of *Pholetesor* was highly susceptible to the traditional organophosphate, carbamate, and synthetic pyrethroid insecticides commonly used in apple orchards.

Tamarixia radiata Waterston [Hymenoptera: Eulophidae], an ectoparasitoid of *Diaphorina citri* Kuwayama [Hemiptera: Psyllidae]

Tamarixia radiata is a species specific ectoparasitoid of the Asian citrus psyllid (ACP) *Diaphorina citri*. *Diaphorina citri* is an economically important pest of citrus in many citrus growing regions of the world. The psyllid vectors the phloem limited gram-negative bacterium *Candidatus* Liberibacter asiaticus, one of the causal organisms of the devastating citrus disease "huanglongbing" (HLB) or citrus greening disease. HLB is responsible for the decline of most trees seen in disease affected regions of the world.

In the United States, ACP and HLB were first identified from Florida in 1998 and 2005, respectively. The psyllid is now well established in citrus producing regions of the state and is also present in Texas, Louisiana, Mississippi, Georgia, South Carolina, Hawaii and southern California. The disease now occurs throughout Florida with highest incidence in the east coast and southwest regions and threatens a citrus industry valued at $1.3 billion in direct sales alone. Other than plant movement by people, *D. citri* is the primary cause of the spread of the HLB disease; therefore, effective and sustainable pest and disease management strategies are required to reduce spread.

Biological control is an essential component of integrated pest and disease management. *Tamarixia radiata* is the most common the two well known primary parasitoids of *D. citri* and thus a strong candidate for classical biological control of this pest. *Tamarixia radiata* was first imported to Florida from Taiwan and south Vietnam and limited numbers were released soon after psyllid was discovered in the state. Evaluation of parasitism of *D. citri* by *T. radiata* at 28 citrus groves across 16 counties in 2006-2007 showed that parasitism rates were generally low, rarely exceeding 20% in contrast to other areas such as Caribbean and Réunion Islands. Therefore, we began importing new parasitoids in 2008 from China, Pakistan and north Vietnam and established colonies of both previously established and new parasitoids which are being released and evaluated in Florida. Other programs of release and evaluation are being conducted in Texas, Mexico and Costa Rica.

Origin

Tamarixia (=Tetrasticus) radiata was first described from northwestern India (Punjab) which is now part of Pakistan.

Appearance

Adults are small wasps that range from 0.92 to 1.04 mm in length. Males are slightly smaller than females both in body length and wing span. Antennae are geniculate and can be used to separate males and females. Male antennae are 1.5 times longer than those of female and contain long and slightly curved hairs or setae compared to female antennae with short setae. Head and thorax are shiny black in both genders. The posterior dorsal and lateral sections of the gaster are black while the venter and a patch of the anterior dorsal gaster are clear to yellow.

Habitat

Tamarixia radiata was recovered throughout the year from fourth and fifth instar nymphs of *D. citri* collected from 26 citrus groves in the central, southwest, and eastern coastal regions of Florida. However, parasitism rates were variable and averaged <20% during spring and summer over all locations. Incidence of parasitism increased during fall at some locations, averaging 39% in September and 56% in November in the central and southwest regions, respectively. Overall, parasitism rates were higher in the southwest region (21%) followed by the central (10%) and eastern coastal (8-12%) regions. Fall is relatively cool and dry compared to hot and humid spring and summer and temperatures are relatively higher in the south than in the north. Parasitism rates from Puerto Rico were much higher and ranged from 79 to 88% between January and April. The relatively more extreme climate of Florida compared with Puerto Rico may reduce parasitoid populations. However, other factors such as high levels of predation, use of insecticides, poor overwintering, or inherent biological characteristics of the parasitoid may be responsible for low parasitism rates in Florida.

PESTS ATTACKED

Diaphorina citri is the only known host of *T. radiata*

Life Cycle

The female usually deposits one and very rarely two eggs on the ventral side of the *D. citri* nymph, mostly in the groove between thorax and abdomen and close to the point of attachment of hind coxae. The neonate larva sucks fluid from the site of attachment and later on moves to neighboring regions as it consumes more body contents and develops underneath the nymph. The parasitized nymph remain indistinguishable from unparasitized nymph for couple of days until almost all body contents of the parasitized nymph are consumed and only wing buds and dorsal covering are left. At that point, the larva is mature and attaches itself to the leaf or shoot surface with a network of fine anchoring silken threads which are later visible around the margins of the mummified nymph. Pupation takes place underneath the mummified nymph and the adult emerges by chewing a circular exit hole in the integument of the thoracic region. Development rate varies with temperature. It takes about 12 days to go from egg to adult at 25 ºC and 14L:10D. A single female can lay up to 300 eggs at 25-30 ºC.

Relative Effectiveness and Conservation

Females of *T. radiata* host feed on the younger instars of *D. citri* and prefer to oviposit underneath the later instars, particularly fifth instar. Through combined behaviors of host feeding and oviposition, a single female is capable of destroying 500 *D. citri* nymphs during her lifetime. *Tamarixia radiata* killed 90% of the presented nymphs through parasitism and host feeding in laboratory experiments. *Tamarixia radiata* was released in Réunion, Taiwan, and Guadaloup, and was credited with reducing the populations of *D. citri* sufficiently in Réunion to mitigate the impact of citrus greening disease. The parasitoid was also detected in Brazil, and Puerto Rico, where no known releases were made. Incidence of parasitism at Isabela, Puerto Rico ranged from 79 to 88% between January and April. High rates of parasitism in spring were followed by continuously reduced psyllid populations during summer. Parasitism rates generally exceeded 50% and averaged 70% at Isabela, but were more variable and averaged 38.5% at Corozál and Gurabo when psyllid populations were lower and more variable.

Tamarixia radiata was imported from Taiwan and south Vietnam, and was released in Florida in 1999. Approximately 37,000 adults of a mixed colony from the two origins were released in Florida citrus between 1999 and 2001. In a study conducted during 2006-2007, *T. radiata* was recovered from fourth and fifth-instar psyllid nymphs at 26 of the 28 groves distributed across the state. However, parasitism rates were much lower compared to other regions such as Réunion, Guadaloupe and Puerto Rico.

Attempts are being made to enhance biological control of *D. citri* by augmenting populations of *T. radiata* with mass-reared *T. radiata* of the already established strain originally imported from Taiwan and south Vietnam and new colonies that we imported from Pakistan, China and north Vietnam. The colony of already introduced strain was established at UF-Southwest Florida Research and Education Center, Immokalee, FL, and colonies from Pakistan, China and north Vietnam populations were established at Division of Plant Industry, Gainesville, FL. Releases of the already established strain and the new strains began in Spring and Fall of 2009, respectively. These parasitoids are being released in both organic and conventional citrus groves and seem to increase the parasitism rates compared to control groves where no releases are being made. For example, in Fall 2009 parasitism averaged 50%

Conservation

We have demonstrated that when using foliar insecticides, more psyllid control is achieved by targeting adults either during the dormant winter period when citrus trees are not flushing or between the flush cycles during the growing season. Such an approach conserves many psyllid natural enemies_that are largely absent from citrus groves during the dormant season but migrate in when trees are producing new growth which is essential for reproduction of psyllids and other prey such as aphids. However, *T. radiata* must also overwinter in citrus groves and is likely to be impacted by dormant sprays. Therefore, it is important to repopulate the groves in spring when psyllid nymphs are available to be parasitized. Psyllid control during the growing season should be based on scouting to further reduce the use of insecticides and conserve beneficials.

Pesticide Susceptibility

Sixteen pesticides including two fungicides were evaluated for toxicity to adult *T. radiata*. Percentage mortality data were evaluated to generally assess IPM compatibility of the pesticides with adult parasitoids. Carbaryl, chlorpyrifos, and fenpropathrin were found to be least compatible with (most toxic to) adult *T. radiata* based on the toxicity of direct sprays and potential long residual life on leaves. Although highly toxic to the parasitoid as direct sprays or freshly dried residues, each of the following was more compatible with *T. radiata* because their residual toxicity was either low at one to three days after application or relatively non-persistent: abamectin, chenopodium oil, fenpyroximate, and spirotetramat. Depending on environmental conditions, imidacloprid (foliar-applied), phosmet, pyridaben, sulfur and 435 spray oil might also be somewhat more compatible for the same reasons. The pesticides that consistently appeared to be most compatible with *T. radiata* were fosetyl-aluminum, copper hydroxide, diflubenzuron, and kaolin clay (Surround WP).

Trichogramma ostriniae Pang et Chen (Hymenoptera: Trichogrammatidae)

As part of a continuing effort to establish natural enemies of the European corn borer, *Ostrinia nubilalis*, in the USA, the egg parasitoid, *Trichogramma ostriniae* Pang et Chen (Hymenoptera: Trichogrammatidae), an important native parasitoid of the Asian corn borer, *Ostrinia furnacalis* (Guenée) was selected for introduction from China. This was accomplished through efforts of entomologists at the University of Massachusetts and USDA-APHIS. Although naturally occurring populations of *T. minutum* Riley, *T. pretiosum* Riley and *T. nubilale* Ertle and Davis, have been reported to parasitize European corn borer eggs on corn in the USA, they generally contribute little to control. In an effort to establish *T. ostriniae* in the northeastern USA, several million female *T. ostriniae* were released in New York from 1991 to 1998. Additional releases were made in Massachusetts, Delaware and other locations. Although, European corn borer egg masses were collected each spring following summers of release, no *T. ostriniae* have been recovered thus far in New York, indicating that it does not overwinter.

Because of its effectiveness against *O. furnacalis* in Asia, *Trichogramma ostriniae* was also identified as a possible candidate for augmentative biological control of European corn borer in the U.S.

Appearance

Like others species of *Trichogramma*, the adult *T. ostriniae* is minute (<0.5 mm in length). The body is relatively compact and the antennae are short. This species is light tan in color and the eyes are red. Host eggs that have been parasitized by *Trichogramma* will turn solid black after about four days.

Habitat

Because this species apparently does not overwinter in New York, it is only found during the season in which it is being released. Inundative and inoculative releases have been made exclusively in sweet corn and this is the only habitat from which it has been recovered.

Pests Attacked (Host Range)

Most effort to date with *T. ostriniae* has been directed against European corn borer in sweet corn. There have also been limited trials against diamondback moth, *Plutella xylostella* in crucifers and against European corn borer in peppers. These are the only two pests known to be attacked under field conditions.

In laboratory studies, eggs of 13 species of Lepidoptera were parasitized by *T. ostriniae*. Eggs of the Noctuidae, Pyralidae, and Plutellidae experienced higher levels of parasitism than others tested.

Female *T. ostriniae* seek out and parasitize host eggs. More than one egg may be inserted into each host egg and this is based, at least in part, on the egg size. After hatching, the parasitoid larvae feed on the contents of the host egg. The wasps pupate within the egg and adults chew an emergence hole to escape. At a constant 80°F, it takes about 10 days from the time of parasitism to emergence of wasps. Under laboratory conditions, on average, 2.1 wasps emerged from eggs of the corn earworm, *Helicoverpa zea*, vs 1.0, from the smaller diamondback moth egg.

In the absence of food, female wasps live about 2.7 days and produce 22 progeny. If they have access to honey, their lifespan increases to 13.9 days and the number of progeny increases to 86. This implies that the provision of food (e.g., flowers with nectaries) under field conditions may improve the effectiveness of releases.

Typically, 80% of the progeny are female. *T. ostriniae* can successfully parasitize eggs of the European corn borer until the "blackhead" stage (larvae are within 24 hrs of emerging). This relatively broad window of susceptibility to parasitism increases the likelihood of successful attack and control in an augmentative biological control program.

Relative Effectiveness

Trichogramma ostriniae has been shown to be effective against the Asian corn borer, *O. furnacalis*.

In the first field trials with *T. ostriniae* in the USA, Mason et al. (unpublished), recorded 97.3% parasitism of naturally occurring European corn borer eggs in sweet corn. Additional inundative release studies have shown that this species is capable of parasitizing high levels (>80%) of European corn borer egg masses. However, despite the high levels of parasitism, damage to sweet corn was still too high.

More recent work has demonstrated that early season inoculative releases show promise. Releases of 30,000 *T. ostriniae* per acre have resulted in season-long parasitism of corn borer egg masses. The relatively few released wasps, successfully reproduced and dispersed in sweet corn. In 1997, we recorded parasitism levels of 50-80% two months following an inoculative release, even in fields treated with insecticides for control of worm pests. Results in 1998 were similar to those in 1997. Whether or not these releases reduced damage to the crop is not known.

Researchers reported that releases of *T. ostriniae* made in sweet corn in Massachusetts, were adversely affected by high or low temperatures and that more European corn borer eggs were parasitized in the lower two-thirds of the sweet corn plant than in the upper one-third. They also reported that parasitism was similar on day two or three following a release, suggesting a limited life span for wasps in the field. If *T. ostriniae* is to be used for inundative release, they recommended that releases should be made every two to three days. Researchers have also investigated the functional response of *T. ostriniae* to various densities of European corn borer egg masses.

In contrast to the recent research with inoculative releases in New York (discussed above), the results reported by the researchers suggest that *T. ostriniae* does not persist very long in the field. It is possible that the persistence observed in New York is due to this species now overwintering and not due to the inoculative releases. Additional study is required to determine if *T. ostriniae* is now successfully overwintering.

Conservation

If *Trichogramma* is being released, the use of broad-spectrum insecticides should be minimized. Pest-selective insecticides would be preferred.

The conservation of natural enemies is probably the most important and readily available biological control practice available to growers. Natural enemies occur in all production systems, from the backyard garden to the commercial field. They are adapted to the local environment and to the target pest, and their conservation is generally simple and cost-effective. With relatively little effort the activity of these natural enemies can be observed. Lacewings, lady beetles, hover fly larvae, and parasitized aphid mummies are almost always present in aphid colonies. Fungus-infected adult flies are often common following periods of high humidity. These natural controls are important and need to be conserved and considered when making pest management decisions. In many instances the importance of natural enemies has not been adequately studied or does not become apparent until insecticide use is stopped or reduced. Often the best we can do is to recognize that these factors are present and minimize negative impacts on them. If an insecticide is needed, every effort should be made to use a selective material in a selective manner.Pesticide Susceptibility

Adult wasps are at great risk from insecticide applications and most will be killed by applications of broad-spectrum materials. However, wasps developing within host eggs are somewhat protected from insecticides and will develop and ultimately emerge.

Trissolcus basalis (Hymenoptera: Scelionidae)

The egg parasitoid *Trissolcus basalis* has been present in the southern

U.S. for many decades. It is known worldwide because of its importance as a parasitoid of the southern green stinkbug, *Nezara viridula*.

Appearance

Minute black wasp with downward elbowed antennae and flattened abdomen. Wing veins are not obvious.

Habitat (Crops)

T. basalis most likely occurs in all crops attacked by southern green stinkbug including cotton, grains, soybeans and other legumes, tomatoes and other solanaceous crops, sweet corn, sunflower, cole crops, cucurbits, fruit and nut crops.

Pests Attacked

The primary host of *T. basalis* is the southern green stinkbug. *T. basalis* also attacks the eggs of other species of stinkbugs.

Life Cycle

T. basalis adults mate immediately after emerging from host eggs. The female typically inserts one egg into a host egg. Heaviest parasitoid egg production occurs during the first few days after emergence, then tapers off. The average number of eggs produced per female was 230-300 in one laboratory study. Adults emerge from the host eggs in 9 to 12 days. The life-cycle averages about 23 days at 22°C.

Relative Effectiveness

Southern green stinkbug was accidentally introduced into Australia in 1916, New Zealand in 1944, and Hawaii in 1961. *T. basalis* has long been considered to be effective at controlling southern green stinkbug in these regions. However, recent studies indicate that control by *T. basalis* may not be the success previously thought. In addition, *T. basalis* has not provided consistent control of southern green stinkbug on soybeans in the southern United States.

Strains of *T. basalis* from Italy, Spain, and France were introduced into California in 1987 and field research showed that 87% of the eggs per mass and 80% of all southern green stinkbug egg masses examined by researchers were parasitized by *T. basalis*. *T. basalis* disperses and reproduces well, it parasitizes other stinkbug species when its preferred host is not available, and it is not known to be subject to hyperparasitism or pathogens in North America. *T. basalis* has a high female to male ratio (up to 5:1) which increases its effectiveness as a parasitoid.

Pesticide Susceptibility

In field trials, *T. basalis* was little affected by applications of permethrin, but was susceptible to methyl parathion, especially within the first six hours. The methyl parathion also killed some wasps as they exited the host eggs, probably from spray residue on the egg exterior being ingested as they chewed through.

Conservation

A nectar source is helpful to other scelionid wasps and may also provide nourishment for *T. basalis*.

FLIES

Pseudacteon spp. (Diptera: Phoridae)

Phorid flies in the genus *Pseudacteon* are parasitoids of ants. Several ant genera serve as hosts, including *Crematogaster, Lasius, Linepithema,* and *Solenopsis* (Disney 1994). Here I focus on *Pseudacteon* species that parasitize *Solenopsis* fire ants, as this is the group on which most research has been done. Over 20 *Pseudacteon* species are known to parasitize *Solenopsis saevissima* complex fire ants in South America. Similarly, more than 20 *Pseudacteon* species parasitize *Solenopsis geminata* complex fire ants from North America to northern South America. None of the North or Central American *Pseudacteon* species parasitize the red imported fire ant, *S. invicta,* or the black imported fire ant, *S. richteri,* both accidentally introduced from South America and now major pests in the United States. *Pseudacteon* species from South America that parasitize *S. saevissima* complex ants represent classical biological control agents for imported *Solenopsis* fire ants in the United States.

Appearance

Pseudacteon phorid flies are very small, about the size of their host ant's head. Seen under a microscope, adult *Pseudacteon* flies have relatively large eyes, a sort of humped back ('humpbacked' flies are one of the common names for this family) and in females, an elaborately shaped ovipositor that varies among species. In the field, they appear as minute, fuzzy specks as they hover over host ants.

Habitat

Pseudacteon species that parasitize *Solenopsis* fire ants are widely distributed in the natural range of their hosts. In tropical and subtropical areas, *Pseudacteon* species are active all year. In temperate regions, *Pseudacteon* adults are active in all except the winter months. *Pseudacteon* phorids are not known to be attracted in large numbers to anything other than host ants. Many phorids appear to be generalists, feeding on honeydew, plant sap, nectar, and dead insects, and *Pseudacteon* adults are also likely to be generalist feeders.

Pests Attacked (Host Range)

Individual *Pseudacteon* spp. are almost always restricted to a single ant genus. A given *Pseudacteon* species may attack multiple *Solenopsis* spp., but is usually restricted to a single species complex, and often reveals affinities for a single species. The host specificity of *Pseudacteon* that attack *Solenopsis* has been documented in tests conducted:

(1) in the field in South America,

(2) in the lab prior to the release of South American *Pseudacteon* species to North America, and

(3) in the field in North America after the establishment of introduced populations.

Life Cycle

Pseudacteon are solitary parasitoids. Each female *Pseudacteon* may produce 100 - 300 eggs. The adult female inserts a single egg into a worker ant with a hypodermic-style ovipositor in a rapid aerial attack. The egg is inserted into the thorax region, and the larva migrates to the head capsule of the worker. There are three instars, during which time worker ants appear to behave normally until just before pupariation.

At pupariation, the tissue inside the ant's head capsule is consumed, killing the ant in the process. The ant's head usually falls off, and the mouth parts are pushed away so that the puparium is visible inside the oral cavity. The pupa completes development in the head capsule, and the adult fly emerges from the oral cavity.

Larval development takes 2-3 weeks and pupal development requires an additional 2-3 weeks, depending upon temperature and the *Pseudacteon* species (larger species have longer developmental times). Development in *Pseudacteon* has been studied most thoroughly in *P. litoralis* and *P. tricuspis*. Sex in most *Pseudacteon* species is apparently determined environmentally, with males produced from smaller workers and females produced from larger workers.

Female *Pseudacteon* are attracted to worker ants at disturbed mounds, mating flights, or foraging trails, although host location behavior varies among species. Introduced *P. tricuspis* in Florida were primarily attracted to interspecific interactions involving their hosts. In some *Pseudacteon* species, males are also attracted to host ants where *Pseudacteon* matings occur. In other *Pseudacteon* species, however, males have not been collected in the vicinity of their host ants (although females may be abundant) and mating apparently occurs elsewhere.

Relative Effectiveness

Pseudacteon parasitism rates have been documented to be very low (<3%). In the presence of *Pseudacteon* species, however, *Solenopsis* worker behavior is affected and foraging is disrupted, leading to dramatic short-term decreases in resource retrieval. Although forager numbers may decrease, *Solenopsis* workers do not usually completely abandon food resources in the presence of *Pseudacteon,* but some workers remain behind to guard the resource. Host *Solenopsis* workers have sometimes been observed to lose control of food resources to competing ant species in the presence of *Pseudacteon* flies, although most of the time they appear to remain in control of the resource.

It is unclear to what degree the observed short-term behavioral effects may translate into long-term population-level impacts. A multi-year field experiment in Florida did not find any measurable effect of introduced *P. tricuspis* on *S. invicta* abundance.

Conservation

The conservation of natural enemies is probably the most important and readily available biological control practice available to growers. Natural enemies occur in all production systems, from the backyard garden to the commercial field. They are adapted to the local environment and to the target pest, and their conservation is generally simple and cost-effective. With relatively little effort the activity of these natural enemies can be observed. Lacewings, lady beetles, hover fly larvae, and parasitized aphid mummies are almost always present in aphid colonies. Fungus-infected adult flies are often common following periods of high humidity. These natural controls are important and need to be conserved and considered when making pest management decisions. In many instances the importance of natural enemies has not been adequately studied or does not become apparent until insecticide use is stopped or reduced. Often the best we can do is to recognize that these factors are present and minimize negative impacts on them. If an insecticide is needed, every effort should be made to use a selective material in a selective manner.

Pesticide Susceptibility

Little is known. It has been found that exposure of parasitized ants to methoprene and pyriproxyfen—insect growth regulators that are the active ingredient in some fire ant baits—can reduce *P. tricuspis* emergence.

Trichopoda pennipes (Diptera: Tachinidae)

Although several species of tachinid flies have been imported as biological control agents of agricultural pests, this distinctive-appearing parasitoid occurs naturally through much of the United States. *Trichopoda pennipes* is a parasitoid of several true bugs.

Appearance

The adult fly is about the size of a house fly and may hover above squash plants in search of prey. It has a bright orange abdomen, velvety black head and thorax, and a fringe of short black hairs on the hind legs. The wings of male flies have a dark spot. The tip of the abdomen of female flies is black.

Habitat

Squash and other crops preyed upon by its key hosts.

Pests Attacked (Host Range)

Primarily squash bug and southern green stinkbug. Although *Trichopoda pennipes* is a parasitoid of several true bugs, there appear to be different biotypes across the country. Limited field studies suggest that different populations, or perhaps cryptic species, attack different species (types) of true bugs. For example, in California, a population of *T. pennipes* has been reported attacking the bordered plant bug. However the same fly was never found attacking the squash bug.

Trichopoda pennipes

Recently, *Trichopoda pennipes* were collected from fields of squash in New York and shipped to California where they were released. These flies have permanently established populations in northern California near farms growing summer and winter squash. In the past, nymphs of squash bug in this area were never attacked by parasitoids. Now, one can easily find 50% or more of these nymphs with fly eggs deposited on them.

Life Cycle

T. pennipes overwinters as a larva in its overwintering host and emerges in late spring or early summer. The female fly lays one to many small, white or gray, oval eggs on large numphs or adult bugs. The larvae burrow from the egg directly into the bug's body. Only one larva survives within each pest bug. A large, cream-colored maggot exits from the body of the bug, drops to the ground, and pupates in a dark reddish-brown, capsule-like puparium. The bug soon dies.

A new generation of adult flies emerges to lay eggs about two weeks later. Each female fly may lay several hundred eggs, and there may be three generations each year, depending on location. The parasitoid overwinters as a larva within the body of the overwintering bug, emerging in late spring or early summer.

Relative Effectiveness

Parasitism of squash bugs may be as high as 80%. However, parasitized bugs continue to feed and lay eggs for some time, and the influence of the fly is not always sufficient to prevent economic crop damage.

Conservation

The conservation of natural enemies is probably the most important and readily available biological control practice available to growers. Natural enemies occur in all production systems, from the backyard garden to the commercial field. They are adapted to the local environment and to the target pest, and their conservation is generally simple and cost-effective. With relatively little effort the activity of these natural enemies can be observed. Lacewings, lady beetles, hover fly larvae, and parasitized aphid mummies are almost always present in aphid colonies. Fungus-infected adult flies are often common following periods of high humidity. These natural controls are important and need to be conserved and considered when making pest management decisions. In many instances the importance of natural enemies has not been adequately studied or does not become apparent until insecticide use is stopped or reduced. Often the best we can do is to recognize that these factors are present and minimize negative impacts on them. If an insecticide is needed, every effort should be made to use a selective material in a selective manne

4

Pathogens and Antagonists of Plant Disease and Post-Harvest Decay

PATHOGENS

Insects and mites, like plants, humans, and other animals, can be infected by disease-causing organisms such as bacteria, viruses, and fungi. Under some conditions, such as high humidity or high pest abundance, these naturally occurring organisms may multiply to cause disease outbreaks or epizootics that can decimate an insect population. Diseases can be important natural controls of some insect pests.

Some pathogens have been mass produced and are available in commercial formulations for use in standard spray equipment. These products are frequently referred to as microbial insecticides, biorational, or bio-insecticides. Some of these microbial insecticides are still experimental, others have been available for many years. Formulations of the bacterium, Bacillus thuringiensis or Bt, for example, are widely used by gardeners and commercial growers.

Most insect pathogens are relatively specific to certain groups of insects and certain life stages. The microbial products do not directly affect beneficial insects and none are toxic to wildlife or humans. Specificity, ironically, can be a disadvantage to the commercialization of these products because their small market may limit profitability.

Unlike chemical insecticides, microbial insecticides can take longer to kill or debilitate the target pest. This may limit their use to crops that can sustain some insect damage. To be effective, most microbial insecticides must be applied to the correct life stage of the pest, and some understanding of the target pest's life cycle is required. Some microbial insecticides must be eaten by the insect to be effective. Good spray coverage is therefore important.

Major characteristics of insect pathogens

- they kill, reduce reproduction, slow growth, or shorten the life of pests
- they usually are specific to target species or to specific life stages
- their effectiveness may depend on environmental conditions or host abundance
- the degree of control by naturally occurring pathogens may be unpredictable
- they are relatively slow acting; they may take several days or longer to provide adequate control
- they may cause epizootics
- Microbial insecticides are compatible with the use of predators and parasitoids, which may help to spread some pathogens through the pest population. Beneficial insects are not usually affected directly because of the specificity of a microbial product, but some parasitoids may be affected indirectly if parasitized hosts are killed. Insecticide applicators should note that although microbials are non-toxic to humans in the conventional sense, safety precautions should be followed to minimize exposure.

ANTAGONISTS

Antagonists of plant disease and food spoilage microorganisms are not yet well understood. However, the research that has been done has yielded exciting and promising results, and the study of antagonists has become a rapidly expanding field in plant pathology.

Worldwide, diseases of crop plants cause losses estimated to be 12%, and post harvest losses due to food spoilage have been estimated to be between 10% and 50%. In the United States, these figures are estimated to be 12% and 9%, respectively. Finding ways to prevent microorganisms from causing these losses would help ensure a stable food supply for the world's ever expanding population. Outside of agriculture, diseases can cause the destruction of entire stands of plants in marshes, forests, or other natural settings, and in other plant systems.

Knowledge of the interactions among microorganisms and ways to manipulate microbiota is growing as research in this field rapidly expands. Antagonists have been successfully used to suppress tomato mosaic, foot and butt rot of conifers, citrus tristeza disease, and crown gall of several crops. Seeds have been coated with antagonists that reduce infection by pathogens and also enhance plant growth. Brown rot of peaches in storage was controlled under simulated commercial conditions by incorporating the antagonist Bacillus subtilis into wax used in the packing process. Inoculation of hosts with antagonists has been used with good results against a common fungal pathogen of conifers and chestnut blight. The future also holds much promise for the suppression of plant-parasitic nematodes by microbiota.

Growers have applied antagonists to the above-ground parts of plants, to the soil (and roots), and to plant seeds. The above-ground environment is the least stable for antagonists because of the extreme variability in moisture and nutrients. Soil is a more stable environment for microbiota, but soil in most fields is generally nutrient poor, pH may range from 4-8, and temperatures and moisture may vary widely. In contrast, greenhouse planting mixes can be managed more effectively to promote antagonist colonization. Finally, it is practical to treat seeds to favor microbial antagonists.

To be most effective, antagonists of plant disease and food spoilage should be:

- genetically stable
- effective at low concentrations
- easy to culture and amenable to growth on an inexpensive medium
- effective against a wide range of pathogens in a variety of systems
- prepared in an easily distributable form
- non-toxic to humans
- resistant to pesticides
- compatible with other treatments (physical and chemical)
- non-pathogenic against the host plant

Relative Effectiveness

Under ideals conditions, such as in the laboratory, antagonists can completely protect plants from pathogens. In the field, disease control is likely to be less successful.

Proper deployment of the antagonist appears to be crucial. Critical factors include moisture and nutrient availability and pH. If the deployment system can meet the needs of the antagonist, successful colonization is more likely. Careful selection of an aggressive strain of the antagonist is also important.

BACTERIA

Over 90 species of naturally occurring, insect-specific (entomopathogenic) bacteria have been isolated from insects, plants, and the soil, but only a few have been studied intensively. Much attention has been given to Bacillus thuringiensis, a species that has been developed as a microbial insecticide.

- Primary hosts: caterpillars; some beetle and fly larvae
- Key characters: larvae stop eating, become limp and shrunken, die and decompose
- Crops: many
- Commercially available: yes — several species and varieties

Bacillus thuringiensis (Bt) occurs naturally in the soil and on plants. Different varieties of this bacterium produce a crystal protein that is toxic to specific groups of insects. Bt has been available in North America as a commercial microbial insecticide since the 1960s and is sold under various trade names. These products have an excellent safety record and can be used on crops until close to the day of harvest. Bt can be applied using conventional spray equipment but, because the bacteria must be eaten to be effective, good spray coverage is essential.

Habitat (Crops)

Many, including vegetables, cotton, tobacco, tree crops, forest crops, landscaping.

Pests Attacked

Numerous moth and butterfly larvae and some beetle and fly larvae are susceptible to infection. Formulations of Bt variety *kurstaki* are available for the control of many caterpillar pests including imported cabbageworm, cabbage looper, hornworms, European corn borer, cutworms, some armyworms, diamondback moth, spruce budworm, bagworms, tent caterpillars, gypsy moth caterpillars and other forest caterpillars, and Indianmeal moth larvae in stored grain. Less well controlled are corn earworm on corn, codling moth, peach tree and squash vine borers.

Formulations of Bt variety *tenebrionis* and variety *san diego* are registered for use against larvae of Colorado potato beetle larvae and elm leaf beetle adults and larvae.

Bt variety *israelensis* is marketed for use against black flies and mosquitoes, fungus gnats, although unless used on a community-wide basis, it is probably more effective to eliminate standing water and control weeds at the edges of ponds. Bt variety aizawai is used to control wax moth larvae in bee hives and various caterpillars. It is important for control of diamondback moth caterpillar which has developed resistance to Bt variety *kurstaki* in some areas.

Some commercially available Bt varieties and target pests:

- *Bacillus thuringiensis*
- var. *tenebrionis* - Colorado potato beetle and elm leaf beetle larvae
- var. *kurstaki* - caterpillars
- var. *israelensis* - mosquito, black fly, and fungus gnat larvae
- var. *aizawai* - wax moth larvae and various caterpillars, especially the diamondback moth caterpillar

Mode of Action

The toxic crystal Bt protein in commercial formulations is only effective when eaten by insects with a specific (usually alkaline) gut pH and the specific gut membrane structures required to bind the toxin. Not only must the insect have the correct physiology and be at a susceptible stage of development, but the bacterium must be eaten in sufficient quantity. When ingested by a susceptible insect, the protein toxin damages the gut lining, leading to gut paralysis. Affected insects stop feeding and die from the combined effects of starvation and tissue damage. Bt spores do not usually spread to other insects or cause disease outbreaks on their own as occurs with many pathogens.

Action of Bacillus thuringiensis var. kurstaki on caterpillars

Diagram courtesy of Abbott Laboratories

Bt genes have been transferred into other microorganisms to produce more active formulations, some of which are commercially available. Additionally, researchers have genetically engineered varieties of several plant species to express the Bt toxin as part of the plant's normal development. This has led to the production of "insect-resistant" Bt-transformed lines of tobacco, cotton, corn, tomatoes, potatoes, and others. The evaluation and development of management systems for these new plant lines are the subject of considerable research.

Symptoms

Larvae affected by Bt become inactive, stop feeding, and may regurgitate or have watery excrement. The head capsule may appear to be overly large for the body size. The larva becomes flaccid and dies, usually within days or a week. The body contents turn brownish-black as they decompose. Other bacteria may turn the host body red or yellow.

Relative Effectiveness

Some naturally occurring bacteria can cause epizootics, especially if the pest population is under stress from lack of food, overcrowding, or cold weather. These epizootics are not as common as those caused by other naturally occurring pathogens. Commercial formulations of *Bacillus thuringiensis,* however, are widely used. Greenhouses, tree and field crops, waterways and thousands of acres of forests are sprayed annually with commercial Bt products.

Successful use of these Bt formulations requires application to the correct target species at a susceptible stage of development, in the right concentration, at the correct temperature (warm enough for the insects to be actively feeding), and before the insect pests bore into the crop plant or fruit where they are protected. Young larvae are usually most susceptible. Caterpillar growth may be retarded even if less than a lethal dose is eaten. Determining when most of the pest population is at a susceptible stage is key to optimizing the use of this microbial insecticide.

Not all caterpillar pests are equally susceptible to Bt. Beet armyworm has proven difficult to control, and some moth species, including some populations of diamondback moth, a major worldwide pest of cole crops, have evolved resistance to the Bt variety *kurstaki* toxins. Corn earworm, squash vine borer larvae, and codling moth larvae are susceptible, but field control is difficult because they rapidly bore into and are protected by plant tissue. Bt is effective against European corn borer if it is applied just as the larvae are hatching. Bt formulations for use against Colorado potato beetle may vary in effectiveness.

Bt formulations may be deactivated in sunlight and may be effective for only one to three days. Rain or overhead irrigation can also reduce effectiveness by washing Bt from crop foliage. Some formulations, such as those involving the genetic engineering of the Bt toxin, aim to overcome these problems.

Bacillus popilliae and Bacillus lentimorbus

are related, naturally occurring bacteria that have been mass-produced for the control of Japanese beetle larvae in turf since the 1940s. Several commercial products are available. The bacteria, usually applied to the soil, cause "milky disease." Milky disease spores may reproduce within the beetle larvae and establish a resident population capable of causing mortality over several seasons if the soil is sufficiently warm and moist through the summer months. It may take several seasons for the disease to control the pest, and it is preferable to treat a broad area to reduce the impact of immigrating healthy beetles.

Symptoms

Beetle larvae killed by *Bacillus popilliae* and *B. lentimorbus* may turn white, hence the name "milky disease."

BACTERIA

Paenibacillus popilliae

(Eubacteriales: Bacillaceae) formerly *Bacillus popilliae* Milky Disease

Paenibacillus popilliae (formerly *Bacillus popilliae*) is a soil-dwelling, Gram-positive, rod-shaped bacterium. It is responsible for a disease (commonly called **milky spore**) of the white grubs of Japanese beetles.

The Japanese beetle, *Popillia japonica,* was accidentally introduced into the USA early this century. Although it is not a problem in its area of origin, the beetle causes serious damage in the USA. It spread rapidly from the initial sightings in New Jersey (1916) and today it is found over roughly half of the country in almost every state east of the Mississippi. It is a problem as an adult beetle because it feeds on a wide range of ornamental and crop plants, eating the tissues between the veins, and it accumulates on ripening fruit, causing substantial damage. It is also a problem in the larval stage because the adult beetles lay their eggs in turf and the grubs destroy the grass roots.

By the 1930s, the infestation had become so extensive that a search for a control measure was undertaken which led to the discovery in nature of some diseased larvae. Milky spore bacteria were isolated. The term "milky disease" comes from the larva's pure white appearance when infected with *B. popilliae. B. popilliae* was the first insect pathogen to be registered in the U.S. as a microbial control agent.

Resident spores in the soil are swallowed by grubs during their normal pattern of feeding on roots. This ingestion of the spore by the host activates reproduction of the bacteria inside the grub. Within 7–21 days the grub will eventually die and as the grub decomposes, billions of new spores are released into the soil.

Milky Spore in the soil is not harmful to beneficial insects, birds, bees, pets or man; and Milky Spore, like other bacteria, is highly survivable in cold and drought conditions.

Appearance

P. popilliae is a Gram-positive spore-forming rod of1.3 to 5.2 × 0.5 to 0.8 micrometers.

Habitat

A fastidious organism, *P. popilliae* grows only on rich media containing yeast extract, casein hydrolysate, or an equivalent amino acid source, and sugars. Several amino acids are known to be required for growth, as well as the vitamins thiamine and barbituric acid. Trehalose, the sugar found in insect hemolymph, is a favored carbon source, although glucose can also be used.

Pests attacked

Japanese beetle is the exclusive host of the strain of *P. popilliae* which is sold commercially. However, other *P. popilliae* strains (and *P. lentimorbus,* which is considered a strain of *P. popilliae* by some experts) have other scarab hosts and are specific to different beetles in the family Scarabaeidae, which includes the Japanese beetle and the chafers - important pasture pests, but also the beneficial dung beetles.

Spores which reside in the soil and have been ingested by beetle larvae germinate in the larva's gut within 2 days and the vegetative cells proliferate, attaining maximum numbers within 3 to 5 days. By this time, some of the cells have penetrated the gut wall and have begun to grow in the hemolymph, where large numbers of cells develop by day 5 to 10. A few spores also are formed at this stage, but the main phase of sporulation occurs later and is completed by 14 to 21 days when the larva develops the typical milky appearance.

In laboratory conditions, the larva remains alive until this stage and usually contains about 5 × 109 spores. In field conditions, however, there are reports that larvae sometimes die earlier, before the main phase of sporulation is completed. This is of concern because sporulation stops when the host dies and the larva ultimately releases fewer spores to maintain the level of infestation of a site.

Relative effectiveness

The advantages of using commercial preparations of *P. popilliae* include the very narrow host range (they are effective against Japanese beetles, only), their complete safety for man and other vertebrates, their compatibility with other control agents including chemical insecticides and insect-pathogenic nematodes, and their persistence.

The disadvantages include the high cost of production *in vivo,* the slow rate of action, the lack of effect on adult Japanese beetles, which also cause obvious and distressing damage, and the need for large areas to be treated for effect. The narrow host range, which is environmentally very desirable, is also a disadvantage: managers must accurately identify the infesting grub species to determine if it is Japanese beetle. If there are other grubs present, they will not be attacked.

The treatment is most effective when applied on a region- or state-wide basis (or at least to relatively large areas) to reduce overall the levels of beetle infestation. It is less appropriate for use by small landowners, who may control the larvae in their own turf only to find their trees and shrubs being eaten by beetles from their neighbors' properties. Also, because *P. popilliae* is obligately dependent on its hosts for sporulation and because some larvae may not ingest spores (or not ingest enough to cause disease) a periodic resurgence and decline of the pest problem can be expected. The success of the control program must be judged not on this basis but by the fact that over a number of years the mean level of pest damage is lower than it would be in the absence of *P. popilliae*.

Between 1939 and 1953, over 100 tons of spore powder was applied to turf in over 160,000 sites in the U.S. as part of a government program. Larval numbers in the turf decreased 10- to 20-fold and the population stabilized at this new low level with corresponding reductions in the levels of adult beetle damage.

Recent research indicates that in some regions of the U.S., *P. popilliae* appears to be losing its virulence against Japanese beetles. Only 0.2% of larvae collected from field sites showed symptoms of milky disease compared to 1946 with 41.5% disease incidence. Also, a recent field study in Kentucky showed that commercial formulations of *P. popilliae* were only moderately infective (39-44%), that infected grubs consumed the same amount of roots as uninfected grubs, and that lower grub populations could not be linked to infection. Researchers concluded that earlier reports of success were limited to very high infestations of grubs where other stresses may have increased their susceptibility to diseases.

The cause of death in insects infected with *P. popilliae* is not fully known. Physiological starvation caused by the growth of bacterial cells in the hemolymph seems the most likely explanation, and fat reserves of diseased larvae have been shown to be much reduced compared with those of healthy larvae. However, toxins also may be involved because they have been detected in culture filtrates of the bacteria and shown to be lethal on injection. Recently, a crystal protein from sporulating cells of *P. popilliae* was found to have similarities to one of the Cry toxins of *B. thuringiensis*. This protein might contribute to pathogenic invasion through the gut wall.

VIRUSES

- Primary hosts: caterpillars
- Key characters: dead larvae hang as limp, semi-liquified, black, decomposing sacs, sometimes attached to upper leaves
- Crops: many
- Commercially available: for greenhouse grown flowers

Insect-specific viruses can be highly effective natural controls of several caterpillar pests. Different strains of naturally occurring nuclear polyhedrosis virus (NPV) and granulosis virus are present at low levels in many insect populations. Epizootics can occasionally devastate populations of some pests, especially when insect numbers are high.

Insect viruses need to be eaten by an insect to cause infection but may also spread from insect to insect during mating or egg laying. In some cases, for example while searching for suitable hosts for egg laying, beneficial insects such as parasitoids may physically spread a virus through the pest population.

No threat to humans or wildlife is posed by insect viruses. Virus diseases of caterpillar pests may cause indirect mortality of some beneficial larval parasitoids if the host insects die before the parasitoids have completed development. Predators and adult parasitoids are not directly affected. Viruses can overwinter in the environment or in overwintering insects to re-establish infection in subsequent seasons.

The successful commercialization of insect-pathogenic viruses has been limited. Thus far, NPV strains have only been mass produced in living insects, a costly procedure. Viral insecticide development is further hindered by the fact that the viruses are specific to one species or genus, ensuring a relatively small market.

Nuclear polyhedrosis viruses have been registered in the United States for use against cotton bollworm, Douglas-fir tussock moth, gypsy moth, European pine sawfly, beet armyworm, and alfalfa looper, although none of these is being produced commercially. The U.S. Forest Service and USDA ARS produce and use an NPV for gypsy moth, but do not sell it. The only commercially available NPV in the United States is used primarily in the Netherlands for greenhouse grown flowers.

Habitat (Crops)

Most crops and habitats affected by caterpillar pests.

Pests Attacked

Naturally occurring viruses may affect many caterpillar pests. NPV affects alfalfa looper, corn earworm, imported cabbageworm, cabbage looper, cotton bollworm, cotton leafworm, tobacco budworm, armyworms, European corn borer, almond moth, spruce budworm, Douglas fir tussock moth, pine sawfly and gypsy moth. Preparations of granulosis virus have been isolated from several caterpillar species, including imported cabbageworm, cabbage looper, armyworm, fall webworm, and mosquitoes, among many others.

Mode of Action

Viruses invade an insect's body via the gut. They replicate in many tissues and can disrupt components of an insect's physiology, interfering with feeding, egg laying, and movement.

Symptoms

Different viruses cause different symptoms. NPV-infected larvae may initially turn white and granular or very dark. Some may climb to the top of the crop canopy, stop feeding, become limp, and hang from the upper leaves or stems, hence the common name "caterpillar wilt" or "tree top" disease. Victims of a granulosis virus may turn milky white and stop feeding. In both cases, the body contents of the dead larvae are liquified and the cuticle ruptures easily to release infectious viral particles. Death from a virus infection usually occurs within three to eight days.

Relative Effectiveness

A naturally occurring viral epizootic can seriously deplete a pest population. Transmission of the virus through the population may take days or weeks but, if conditions are suitable, the entire population may eventually collapse. In some instances, the combination of naturally occurring viruses and other natural enemies will maintain pest populations at acceptable levels. For example, a virus has been recorded as destroying up to 28% of imported cabbageworm populations in cole crops, with up to 55% of the remaining population parasitized by several parasitoids. Cabbage looper populations experienced up to 40% infection by virus in the same studies.

Mass reared viruses have been successfully applied in limited areas as microbial insecticides against pests. In trials in Maryland, releases of granulosis virus proved as effective as Bt against some caterpillars and, in trials against imported cabbageworm on cole crops in New York, a granulosis virus yielded results equivalent to applications of chemical insecticide. Pheromones (insect attractants) have been used to lure male corn earworm moths to NPV-contaminated bait stations to help spread the virus within a corn earworm population.

Infected caterpillars have been mashed into a water solution and applied to pest populations as a form of microbial insecticide. Viruses are adversely affected by ultraviolet radiation and are best applied in the late afternoon. Pest abundance and the general fitness of the pest population will affect its susceptibility to virus attack, and the effectiveness of different strains of the same virus can vary considerably. Viruses can be stored frozen for many years. Research to reduce the killing time of viruses is ongoing and includes the use of genetic engineering to improve the performance of viral insecticides.

Baculoviruses (Baculoviridae)

Baculoviruses are pathogens that attack insects and other arthropods. Like some human viruses, they are usually extremely small (less than a thousandth of a millimeter across), and are composed primarily of double-stranded DNA that codes for genes needed for virus establishment and reproduction. Because this genetic material is easily destroyed by exposure to sunlight or by conditions in the host's gut, an infective baculovirus particle (*virion*) is protected by protein coat called a *polyhedron* (plural *polyhedra*). Most insect baculoviruses must be eaten by the host to produce an infection, which is typically fatal to the insect.

The majority of baculoviruses used as biological control agents are in the genus *Nucleopolyhedrovirus*, so "baculovirus" or "virus" will hereafter refer to nucleopolyhedroviruses. These viruses are excellent candidates for species-specific, narrow spectrum insecticidal applications. They have been shown to have no negative impacts on plants, mammals, birds, fish, or even on non-target insects. This is especially desirable when beneficial insects are being conserved to aid in an overall IPM program, or when an ecologically sensitive area is being treated. The USDA Forest Service currently uses the gypsy moth nuclear polyhedrosis virus (*LdNPV*) to aerially spray thousands of acres of forest each year. This product, registered as **Gypchek**, is effective against gypsy moths but leaves all other animals unharmed (Reardon et al. 1996).

On the other hand, the high specificity of baculoviruses is also cited as a weakness for agricultural uses, since growers may want one product to use against a variety of pests. Currently, researchers are attempting to use genetic engineering techniques to expand virus host ranges to the desired pest species. Releases of such genetically-engineered baculoviruses have been made by researchers in the U.K. and the United States and show promise, although the cost of commercial production of these agents must be reduced if they are to be competitive. Companies like Dupont, biosys, American Cyanamid, and Agrivirion (to name a few) have continued to explore the expansion and development of agricultural-use viral insecticides.

Recently, biosys has released two baculovirus-based products, Spod-X for beet armyworm and Gemstar LC for tobacco budworm and cotton bollworm.

Life Cycle

The baculovirus life cycle involves two distinct forms of virus. Occlusion derived virus (ODV) is present in a protein matrix (polyhedrin or granulin) and is responsible for the primary infection of the host while the budded virus (BV) is released from the infected host cells later during the secondary infection.

Typically, the initial infection occurs when a susceptible host insect feeds on plants that are contaminated with the occluded form of the virus. The protein matrix dissolves in the alkaline environment of the host midgut (stomach), releasing ODV that then fuse to the columnar epithelial cell membrane of the host intestine and are taken into the cell in endosomes. Nucleocapsids escape from the endosomes and are transported to nucleus. This step is possibly mediated by actin filaments. Viral transcription and replication occur in the cell nucleus and new BV particles are budded out from the basolateral side to spread the infection systemically. During budding, BV acquires a loosely fitting host cell membrane with expressed and displayed viral glycoproteins.

Baculovirus infection can be divided to three distinct phases, early (0–6 h post-infection), late (6–24 h p.i.) and very late phase (18–24 to 72 h p.i.). While BV is produced in the late phase, the ODV form is produced in the very late phase acquiring the envelope from host cell nucleus and embedded in the matrix of occlusion body protein. These occlusion bodies are released when cells lyse to further spread baculovirus infection to next host. The extensive lysis of cells frequently causes the host insect to literally disintegrate, thus the reason for the historic name "wilting disease." the complete ODV-polyhedrin particles are resistant to heat and light inactivation, whereas the naked BV virion is more sensitive to environment.

Relative Effectiveness

It is widely acknowledged that baculoviruses can be as effective as chemical pesticides in controlling specific insect pests. However, the expense of treating a hectare of land with a baculovirus product invariably costs more than an equally efficacious chemical treament. This difference in price is due primarily to the labor intensive nature of baculovirus production. Some viruses can be produced *in vitro* (within cell cultures in the laboratory, not requiring whole, living insects). These are less expensive than those that can only be produced *in vivo,* that is, inside of living insects. The cost of rearing live hosts adds greatly to the final cost of the product. It is to be hoped that insect cell culture systems currently being developed for other uses may ultimately make viral pesticides more cost-effective.

Appearance

Insects killed by baculoviruses have a characteristic shiny-oily appearance, and are often seen hanging limply from vegetation (Fig.E). They are extremely fragile to the touch, rupturing to release fluid filled with infective virus particles. This tendency to remain attached to foliage and then rupture is an important aspect of the virus life-cycle. As discussed above, infection of other insects will only occur if they eat foliage that has been contaminated by virus-killed larvae.

It is interesting to note that most baculoviruses, unlike many other viruses, can be seen with a light microscope. The polyhedra of many viruses look like clear, irregular crystals of salt or sand when viewed at 400x or 1000x. The fluid inside a dead insect is composed largely of virus polyhedra - many billions are produced inside of one cadaver.

Habitat

Baculoviruses can be found wherever insects exist. Because rain and wind readily carry baculoviruses from place to place, it is likely that every piece of land and body of water contains some virus particles. It is widely accepted by researchers that most produce currently on the shelves is "contaminated" by baculovirus particles (Heimpel et al., 1973). In fact, the pervasiveness of baculovirus particles, along with the results of tests performed in conjunction with registration, may be considered both indirect and direct evidence for the safety of these agents.

Pests Attacked

Like most viruses, baculoviruses tend to be species or genus specific, although there are some exceptions to this rule, notably the *Autographa californica* nuclear polyhedrosis virus. Much of the genetics work currently being done to improve baculovirus-based pesticides is concentrated in the area of the virus genome controlling its host range.

Pesticide Compatibility

Viruses particles *per se* are generally unaffected by pesticides, although some chlorine compounds should be expected to damage or destroy viruses if applied at the same time. Baculovirus *efficacy*, however, can be altered in many ways by the effects of chemical pesticides on the host insect. A review showed that of 10 pesticide-virus combinations, 9 resulted in an additive effect on insect mortality. However, some of the pesticides included in that review have since been banned, and this study is of limited use at present.

More work is needed to explore the effectiveness of insecticide "cocktails" consisting of enviromentally friendly chemical agents and baculoviruses.

Current use of baculoviruses as biological insecticides

Table 1

Commodity	*Insect Pest*	*Virus Used*	*Virus Product*
Apple, pear, walnut and plum	Codling moth	Codling moth granulosis virus	Cyd-Xe
Cabbage, tomatoes, cotton,	Cabbage moth, American bollworm, diamondback moth, potato tuber moth, and grape berry moth	Cabbage army worm nuclear polyhedrosis virus	Mamestrin
Cotton, corn, tomatoes	*Spodoptera littoralis*	*Spodoptera littoralis* nuclear polyhedrosis virus	Spodopterin
Cotton and vegetables	Tobacco budworm *Helicoverpa zea*, and Cotton bollworm *Heliothis virescens*	*Helicoverpa zea* nuclear poly-hedrosis virus	Gemstar LC, Biotrol, Elcar
Vegetable crops, greenhouse flowers	Beet armyworm (*Spodoptera exigua*)	*Spodoptera exigua* nuclear polyhedrosis virus	Spod-X
Vegetables	Celery looper (*Anagrapha falcifera*)	Anagrapha falcifera nuclear polyhedrosis virus	none at present
Alfalfa and other crops	Alfalfa looper (*Autographa californica*) virus	*Autographa californica* nuclear polyhedrosis	Gusano Biological Pesticide
Forest Habitat, Lumber	Douglas fir tussock moth (*Orgyia psuedotsugata*)	*Orgyia psuedotsugata* nuclear polyhed-rosis virus	TM Biocontrol
Forest Habitat, Lumber	Gypsy moth (*Lymantria dispar*)	*Lymantria dispar* nuclear polyhedrosis virus	Gypchek

Fungi

- Primary hosts: aphids, whiteflies, leafhoppers, flies, beetles, caterpillars, thrips, and mites; some beetle larvae
- Key characters: reduced feeding, lethargy; dead insects swell, and may be covered with fungi
- Crops: many
- Commercially available: several, including antagonists of fungal disease and a product (*Metarhizium anisopliae*) which is used against household cockroaches

Some insect species, including many pests, are particularly susceptible to infection by naturally occurring, insect-pathogenic fungi. These fungi are very specific to insects, often to particular species, and do not infect animals or plants. Fungal growth is favoured by moist conditions but fungi also have resistant stages that maintain infection potential under dry conditions. Fungi have considerable epizootic potential and can spread quickly through an insect population and cause its collapse. Because fungi penetrate the insect body, they can infect sucking insects such as aphids and whiteflies that are not susceptible to bacteria and viruses.

Several fungal species have potential as microbial insecticides and, in some countries, are commercially available in formulations that can be applied using conventional spray equipment.

Habitat (Crops)

Most crops including soybeans, greenhouse crops, vegetables, cotton, citrus, and ornamentals; also interior plantscapes and forests. An aquatic fungus infects mosquito larvae of some genera.

Pests Attacked

Most insect pests are susceptible to fungal pathogens. Some fungi, such as the *Entomophthora* and related species, are fairly specific with regard to the groups of insects affected; others, such as *Beauveria,* have a wider host range.

Mode of Action

Fungi invade insects by penetrating their cuticle or "skin." Once inside the insect, the fungus rapidly multiplies throughout the body. Death is caused by tissue destruction and, occasionally, by toxins produced by the fungus. The fungus frequently emerges from the insect's body to produce spores that, when spread by wind, rain, or contact with other insects, can spread infection.

Symptoms

Infected insects stop feeding and become lethargic. They may die relatively rapidly, sometimes in an upright position still attached to a leaf or stem, perhaps in an elevated location or concentrated near crop borders. The dead insect's body may be firm and "cheese-like" or an empty shell, often but not always with cream, green, red, or brown fungal growth, either enveloping the body or emerging from joints and body segments. Infected aphids can be swollen and discolored; infected root maggot flies may be clustered on shoot tips, tall grasses, or other prominent features.

Relative Effectiveness

Insect-pathogenic fungi usually need moisture to enable infection, and natural epizootics are most common during wet or humid conditions. The effectiveness of these fungi against pest insects depends on having the correct fungal species and strain with the susceptible insect life stage, at the appropriate humidity, soil texture (to reach ground-dwelling pest species), and temperature. The fungal spores, which can be carried by wind or water, must contact the pest insect to cause infection. Naturally occurring fungal epizootics may decimate aphid, root maggot fly, caterpillar, leafhopper, and thrips populations. They can be an important natural control of aphids in potatoes and other crops.

Many insect-pathogenic fungi occur in the soil. There is evidence that application of some soil insecticides, fungicides, and herbicides can inhibit or kill these fungi. For example, even quite low concentrations of some herbicides can severely limit the germination and growth of *Beauveria bassiana* fungal spores in soil samples.

SOME COMMON INSECT-PATHOGENIC FUNGI

Entomophthora muscae infects flies. Susceptible pest species include the adults of the onion maggot, cabbage maggot, and seedcorn maggot. The fungus multiplies within the body of the adult fly which becomes enlarged; yellowish bands of fungal spores stripe the abdomen.

Related species include *Pandora neoaphidis,* a common, naturally occurring pathogen of aphids that can be particularly effective during moist periods. *Zoophthora radicans* is another common fungus, with a broad host range. Strains have been isolated from several caterpillars, including diamondback moth, leafhoppers, aphids, and some weevils. Other related species infect thrips.

Beauveria bassiana is an insect-pathogenic fungus found naturally on some plants and in the soil. Epizootics are favored by warm, humid weather. It is known as the white muscardine fungus because infected insect larvae eventually turn white or gray. *Beauveria* is used as a fungal microbial insecticide in some countries. It has an extensive host list that includes such important pests as whiteflies, aphids, grasshoppers, termites, Colorado potato beetle, Mexican bean beetle, Japanese beetle, boll weevil, cereal leaf beetle, bark beetles, lygus bugs, chinch bug, fire ants, European corn borer, codling moth, and Douglas fir tussock moth. It has even been found infecting the lungs of wild rodents, and the nasal passages of humans. Unfortunately, natural enemies, such as lady beetles, are susceptible too. One possible application method that may avoid harming beneficial insects is the use of fungus-contaminated insect baits that are attractive to pest species only. There are many different strains of the fungus that exhibit considerable variation in virulence, pathogenicity and host range. It occurs in the soil as a saprophyte.

Metarhizium spp. is being tested as a natural enemy of corn rootworm, white grubs (scarabs), and some root weevils. It has a very broad host range and is extensively used in Brazil against spittle bugs in sugar cane and alfalfa.

Other species of insect-pathogenic fungi have been tested as microbial insecticides for the control of pests. *Verticillium lecanii* is used in Europe against greenhouse whitefly and thrips and aphids, especially in greenhouse crops. *Neozygites floridana* has been successfully tested against spider mites, and Nomuraea rileyi has activity against green cloverworm, cabbage looper, imported cabbageworm, armyworms, corn earworm, and tobacco budworm. *Hirsutella thompsonii* infects mites and was previously registered for use in the United States.

Entomophaga grylli (Zygomycetes: Entomophthorales)

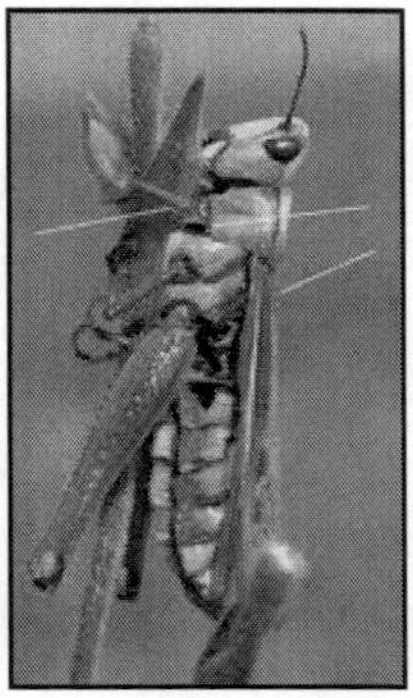

E. grylli

E. grylli is a species complex that has two distinct pathotypes in North America: *Entomophaga macleodii* and *Entomophaga calopteni*. In addition, another complex member, *E. praxibuli* has been introduced from Australia. They all attack grasshoppers which are major pests of crops and rangeland in the Western United States and Canada. In 1986, approximately 55 million acres of western rangeland were infested by grasshoppers and the economic cost is estimated at over $400 million per year.

Appearance

Entomophaga grylli is easily identified in the field. Shortly before death, infected grasshoppers crawl to the top of plants and die with their heads pointing upward and their legs wrapped tightly around the stalks where they may remain for several days.

Habitat

Rangeland and cropland, predominantly in the Western U.S. and Canada.

Pests Attacked (Host Range)

E. macleodii infects grasshoppers in the subfamily Oedipodinae, including band-wing grasshoppers. *E. calopteni* infects grasshoppers in the genus Melanoplus. *E. praxibuli* infects both Oedipodinae and *Melanoplus* species in the United States.

Life Cycle

The members of this species complex can be separated principally by differences in vegetative growth patterns, the formation or lack of primary conidia, the size of and number of conidial nuclei, isozyme polymorphism profiles, and host and geographic ranges.

E. calopteni is monocyclic forming only overwintering resting spores allowing disease cycles to occur in subsequent years. *E. macleodii* and *E. praxibuli* are polycyclic, characterized by forming asexual conidia and overwintering resting spores. Under proper environmental conditions, (high humidity and rainfall) large numbers of conidia are formed on individual grasshopper cadavers and forcibly discharged, resulting in several additional disease cycles per season and dramatic increases in infection levels.

Relative Effectiveness

Periodic and widespread epizootics of grasshoppers in North America have been attributed to *E. grylli*. In Australia, *E. praxibuli* has caused epizootics in grasshopper populations under fairly dry conditions, a consideration in its introduction into the many arid rangeland areas where grasshoppers are pests.

Presently, the *E. grylli* complex has limitations as a biological control agent. It cannot be mass produced, and it is weather-dependent. In addition, there are over 600 species of grasshopper in North America and it is not feasible to test all of these for susceptibility. However, the introduction of *E. praxibuli* was part of a project involving the study of other biological control agents including viruses, parasitoids and the fungal pathogen, *Beauveria bassiana*. In an integrated approach, the *E. grylli* complex would seem to have a niche.

Entomophaga maimaiga (Entomophthorales: Family)

Entomophaga maimaiga is a Japanese fungus which has shown striking success in managing Gypsy Moth populations in North America.

Entomophaga maimaiga was introduced in the United States from Japan in 1910 and 1911, its 1989 appearance in Connecticut, New Hampshire, Vermont, northeastern Pennsylvania, New Jersey, southeastern New York, and Massachusetts is a mystery: if its presence is due to the original introduction, why wasn't it detected during the years between 1911 and 1989?

The fungus spreads through aerial dispersion of actively ejected asexual spores from cadavers of Gypsy Moth larvae it has killed. It can also be spread unwittingly by humans.

The fungus persists in the top layer of soil as resting spores. These have been shown to persist for at least 11 or 12 years, probably longer.

Appearance

Presence of *E. maimaiga* may be determined by late instar gypsy moth larvae which, when infected with this fungus, die hanging vertically from tree trunks with prolegs extended laterally. The cadavers subsequently fall to the bases of trees.

Hyphal bodies are found within infected insects and spore-producing mycelium can grow on the outside of cadavers but decomposes after conidia are produced and actively ejected from cadaver surfaces. Resting spores are about 30 micrometers in diameter and may be found inside late instars of infected gypsy moths.

Habitat

Trees, both broadleaf and coniferous, of forests and urban and suburban areas.

Pests attacked

Based on lab results, *E. maimaiga* appears to be quite specific to the family that includes gypsy moth, although it can cause low levels of infection in a number of other species.

During a field study, the only cadavers that were found on trees were gypsy moth larvae killed by *E. maimaiga*. At locations where there were active epizootics of *E. maimaiga* occurring in gypsy moth populations, more than 1500 insects of 53 difference species were collected. Of those 1500 individuals, only one individual of a lasiocampid and one individual from the Noctuidae were infected with *E. maimaiga*.

Life cycle

Species in the Entomophthorales have several life stages:

- The conidia are relatively short-lived. Spores are actively discharged from cadavers and can immediately germinate and cause infection.
- Hyphal bodies occur within infected insects and mycelium grows on the outside of cadavers.
- Resting spores are produced internally in late instars and must undergo a period of dormancy.

Relative Effectiveness

From *E. maimaiga* epizootics occurring in North American gypsy moth populations in 1989 and 1990, it became clear that this fungus was capable of becoming an important mortality factor. Whereas nuclear polyhedrosis viral epidemics in gypsy moth populations require a high population density, *E. maimaiga* can cause epizootics at lower densities. In release studies done in 1991 and 1992, this fungus appeared to spread readily, even in less wet years, and persisted 3 years later. Gypsy moths were also still present in these studies, but there was a high degree of fungal infection documenting long-term persistence of this fungus in an area with the potential for long-term control. In a 1995 study of *E. maimaiga* released against gypsy moth in residential areas, only the control plots had high levels of defoliation.

Data from 1996 indicates that gypsy moth defoliation declined more than 85% from 1995 levels over 11 states. The most dramatic case is in Virginia where 1996 defoliation was nonexistent compared with 850,000 acres in 1995. There is general consensus among scientists and pest managers that *E. maimaiga* is probably responsible for the decline. However it is not known if this level of fungus activity will continue. It appears that *E. maimaiga* has spread throughout much of the area considered to be generally infested by gypsy moth, the only exception being the transition area into which the insect is spreading for the first time.

E. maimaiga has been highly variable and unpredictable. Consequently, the use of environmentally safe and effective insecticides will continue to be important tools to reduce damage caused by gypsy moth outbreaks, and several points should be emphasized:

- It is important to continue studies to fully understand how *E. maimaiga* will affect gypsy moth populations in the long term. As a new member of a group of natural control agents that regulate gypsy moth population in North America, it is not known how the fungus will fit into the cycle in the long term or what effect it might have on other natural control agents of the gypsy moth.

- Gypsy moth outbreaks will continue to occur and may initially be severe as the insect spreads into uninfested areas. However, it is hypothesized that as the fungus becomes established in these new areas of gypsy moth infestation it will have a moderating influence on the gypsy moth population.
- People will continue to inadvertently move the gypsy moth to uninfested areas of the country on outdoor household articles, nursery stock, Christmas trees, and other commodities. Detection, monitoring, and suppression of these isolated infestations will continue to be important to prevent establishment outside the generally infested area.

Conservation

Each spring, infection by *E. maimaiga* is initiated when resting spores germinate to produce infective germ conidia. Because high titers of resting spores have been found in the organic layer of the soil at the bases of trees, it is assumed that disruption of this layer, e.g., turning over the soil at the tree base, might decrease the titer of resting spores near the soil surface, and thus would decrease the level of infection from these spores.

Although *E. maimaiga* has not yet been tested for its susceptibility to fungicides, there is the potential that use of fungicides especially around the bases of trees could cause mortality among *E. maimaiga* resting spores.

Pesticide Susceptibility

Studies are planned to evaluate the susceptibility of *E. maimaiga* to the fungicides to which it might be exposed in the forest and urban landscape.

Entomophthora muscae

(=Entomophthora schizophorae)
Zygomycetes: Entomophthorales

Female *Melanostoma scalare* hoverfly

Entomophthora muscae is a species of pathogenic fungus in the order Entomophthorales which causes a fatal disease in flies. It can cause epizootic outbreaks of disease in houseflies and has been investigated as a potential biological control agent.

Entomophthora muscae is a well-known fungal disease of adult

Diptera with a wide range of potential hosts. First described by Cohn in 1855 from an epizootic of house flies, *E. muscae* has been recognized as a potential biological agent for many years.

Appearance

Patent infection in the house fly is characterized by a distended abdomen, legs spread and wings out-stretched (see photograph above). The intersegmental membranes of the swollen abdomen give the cadaver a characteristic striped pattern. The labellum may be lowered and the cadaver may be attached to the surface by the mouthparts. A halo of primary conidia often surrounds the cadaver, and is particularly evident on glass surfaces. The release of primary conidia from the cadaver is termed a "conidial shower." Disease transmission depends on direct contact with these conidia.

Habitat

Entomophthora muscae epizootics usually occur in the spring and fall of the year, and occur in most temperate regions. Sporulation is timed to coincide with periods of cool, and humid conditions in areas with an abundance of resting hosts. Such environmental conditions promote primary conidial production and disease transmission. In the northeastern U.S., for example, *E. muscae*-killed cluster flies are frequently seen attached, indoors, to windows or window frames where other flies tend to aggregate. In agricultural settings, fly cadavers may be seen attached to the underside of plant leaves, along field margins, on fence rails and feed bunks, and in dairy barns and poultry houses.

Pests Attacked (Host Range)

Since Cohn's initial observations, epizootics have been observed in adult Diptera in the families Muscidae, Calliphoridae, Sarcophagidae, Tachinidae, Drosophilidae, Scatophagidae, Culicidae and Syrphidae.

Outbreaks of infection with *Entomophthora muscae* tend to occur in the spring and autumn. The fungus is found in most temperate regions and sporulation usually takes place in cool, humid conditions in areas where flies congregate and rest. In houses, the corpses of flies are frequently seen attached to windows or window frames, a place where other flies may easily become infected. In the open, they may be seen attached to the underneath of leaves, on fences and walls, in agricultural buildings and poultry houses

Life Cycle

There are at least two forms of conidia produced by *E. muscae*. Large primary conidia are produced soon after the fly dies, but in the event no hosts are available for the primary conidia, a smaller secondary conidia may develop from the primary conidia.

Conidia are produced at the apex of a conidiophore, and when mature, are forcibly ejected onto the surrounding surface area. Conidia may fall onto un-infected resting flies close-by. The conidia germinate within a few hours and a germ tube begins to penetrate the insect's cuticle. Once inside the hemocoel, the cytoplasm streams through the germ tube and into the insect's hemolymph. The fungus replicates in the fly hemolymph, and progressively consumes all the hemolymph in about 5-7 days. In the evening hours, critically ill insects tend to crawl up verticle surfaces to die at an elevated position. Approximately 3 hours after death the condiophores begin to emerge from the intersegmental membranes to initiate the conidial shower.

Relative Effectiveness

Epizootics appear to be influenced by abiotic factors, particularly temperature (Watson and Petersen 1993). When mid summer temperature were high the prevalence of disease decreased to near non existent levels, but recovered late in the season with the arrival of cooler field conditions. *Entomophthora muscae* was prevalent on two dairy farms in Nebraska, where 76 and 80% of the house fly population were infected. In laboratory studies, *E. muscae* infected house flies exhibited behavioral fevers and were able to rid themselves of infection by resting at temperatures that inhibited fungal growth.

Conservation

Unfortunately, *E. muscae* is a fastidious organism and cannot be easily cultured artifically. Colonies are usually maintained through direct fly to fly transmissions. Hyphal bodies of *E. muscae* have been stored in liquid nitrogen. However, artificial infection through injection has met with limited results.

Pesticide Susceptibility

Insecticide residues significantly inhibited the germination of primary conidia. Permethrin and naled had the least inhibition followed by tetrachlorvinphos/dichlorvos and carbaryl, and malathion and dimethoate.

Use in Biological Control

There are technical difficulties in controlling flies with *Entomophthora muscae*. The fungus is sensitive to temperature and when the air temperature is high, the prevalence of the disease decreases to very low levels.Houseflies infected with *E. muscae* were able to rid themselves of the infection by resting at temperatures that inhibited the growth of the fungus. Storage of the organism in a viable state is difficult and it cannot be easily cultured in the laboratory. Captive colonies need to be maintained through direct transmission from fly to fly. There is no commercially available supply of the pathogen.

Metarhizium (Order: Hypocreales, Family: Clavicipitaceae)

Metarhizium is a genus of entomopathogenic fungi in the Clavicipitaceae family. With the advent of genetic profiling, it has now become possible to place these fungi in proper taxa. Most turn out to be the asexual forms (anamorphs) of fungi in the phylum Ascomycota.

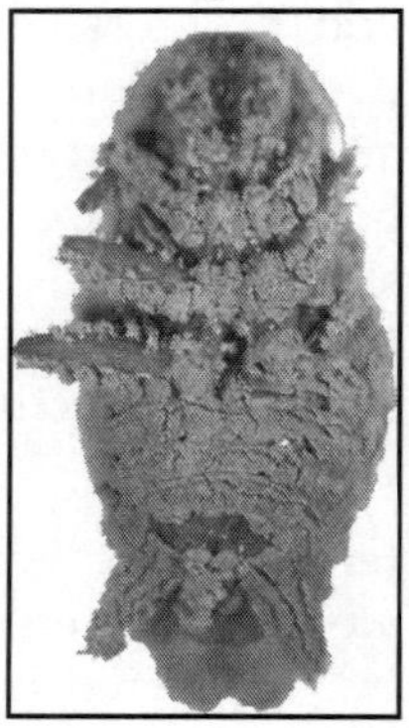

Cockroach killed by *Metarhizium anisopliae*

Metarhizium species (Metschnikoff) Sorokin, also known as green muscardine fungi, have long been recognized for their biological control potential against arthropods. As early as 1879, fungi from this genus were being evaluated for control of wheat chafer beetles, *Anisoplia austriaca,* and sugar beet curculio, *Cleonus punctiventris,* in Ukraine. The species level name of one of the more widely researched *Metarhizium* species (*M. anisopliae*) was derived from this beetle. Morphological features for identifying *Metarhizium* species can be imprecise as there can often be overlap of characters among species. Molecular techniques have shown that what used to be called *M. anisopliae* represents a complex of nine species.

Appearance

Infections of arthropods by *Metarhizium* species are easily recognized a few days after death, when the fungus grows out of the arthropod integument and forms reproductive structures. Initially, one only sees fungal hyphae that appear white, but, as conidia form and mature they often take on a characteristic olive green colour. However, depending on the species and strain of *Metarhizium,* spores can range in color from white to yellow to brown and green.

Habitat (Crops)

Metarhizium species are commonly thought of as soil saprophytes and are most frequently found in disturbed habitats like agricultural fields as compared to forest ecosystems. Additionally, recent findings suggest that these fungi form associations with plant roots in the rhizosphere and survive better in that environment than in surrounding potting soil over extended periods of time.

Pests Attacked

Metarhizium species are known to attack a wide range of arthropods: greater than 200 species in over 50 families. These include many species of agricultural, medical and veterinary importance. Some insect hosts included on two active product labels in the U.S. include "various ticks and beetles; root weevils, flies, gnats, thrips," and locusts and grasshoppers (Green Muscle, Becker Underwood, Ames, Iowa). Additionally, *Metarhizium* species have been developed in other countries for use against cockchafers, spittlebugs, grubs, borers, and for control of mosquitoes that vector malaria.

Life Cycle

Our discussion of the life cycle of species within this genus will be restricted to their activity as arthropod pathogens. Generally, the development of a lethal arthropod infection can be separated into three stages. First, asexual conidia (singular=conidium) come into contact with the arthropod integument as the arthropod travels through the environment. The conidia stick to the arthropod's exoskeleton, germinate and grow a germ tube, which eventually ends in an appressorium, the flattened and thickened tip of a germ tube. A penetration peg grows under the appressorium, pierces the integument and enters the hemocoel. The penetration of the fungus is achieved by the production of a cocktail of hydrolytic enzymes including proteases, lipases, chitinases, and mechanical pressure. Second, single cells of the fungus, blastospores, bud off of the penetration structure, circulate in the insect hemocoel and multiply, thereby depleting host nutrients. *Metarhizium* species are also known to produce compounds that are toxic to arthropods and presumably aid in killing the host, suppressing host immune defenses and fending off potential microbial competitors. Finally, after the host dies due to mycosis, the fungus will penetrate out of the integument and grow conidiophores, on which environmentally stable aerial conidia are produced. These conidia are passively disseminated into the environment and eventually infect new hosts.

Pesticide Susceptibility

The effect of fungicides on *Metarhizium* species can be variable and needs to be assessed on a case-by-case basis. If there is any doubt as to the effect of a fungicide or insecticide on the viability of *Metarhizium* species products, users should err on the side of caution and not use tank mixes. Bruck (2009) reported on the persistence of *M. anisopliae* conidia in Petri dishes, bulk soil and in the rhizosphere, when in contact with several fungicides commonly used in greenhouses and nurseries. While a large number of fungicides negatively impacted conidial germination and mycelial growth in Petri dishes, there was little effect in the rhizosphere following two applications of some fungicides. The fungicides with the largest detrimental impact were Cap tan and triflumizole.

NEMATODES (Rhabditida: Steinernematidae & Heterorhabditidae)

Nematodes are simple roundworms. Colorless, unsegmented, and lacking appendages, nematodes may be free-living, predaceous, or parasitic. Many of the parasitic species cause important diseases of plants, animals, and humans. Other species are beneficial in attacking insect pests, mostly sterilizing or otherwise debilitating their hosts. A very few cause insect death but these species tend to be difficult (e.g., tetradomatids) or expensive (e.g. mermithids) to mass produce, have narrow host specificity against pests of minor economic importance, possess modest virulence (e.g., sphaeruliids) or are otherwise poorly suited to exploit for pest control purposes. The only insect-parasitic nematodes possessing an optimal balance of biological control attributes are entomopathogenic or insecticidal nematodes in the genera *Steinernema* and *Heterorhabditis*. These multi-cellular metazoans occupy a biocontrol middle ground between microbial pathogens and predators/ parasitoids, and are invariably lumped with pathogens, presumably because of their symbiotic relationship with bacteria.

Entomopathogenic nematodes are extraordinarily lethal to many important insect pests, yet are safe for plants and animals. This high degree of safety means that unlike chemicals, or even *Bacillus thuringiensis,* nematode applications do not require masks or other safety equipment; and re-entry time, residues, groundwater contamination, chemical trespass, and pollinators are not issues. Most biologicals require days or weeks to kill, yet nematodes, working with their symbiotic bacteria, can kill insects within 24-48 hours. Dozens of different insect pests are susceptible to infection, yet no adverse effects have been shown against beneficial insects or other nontargets in field studies. Nematodes are amenable to mass production and do not require specialized application equipment as they are compatible with standard agrochemical equipment, including various sprayers (e.g., backpack, pressurized, mist, electrostatic, fan, and aerial) and irrigation systems.

Hundreds of researchers representing more than forty countries are working to develop nematodes as biological insecticides. Nematodes have been marketed on every continent except Antarctica for control of insect pests in high-value horticulture, agriculture, and home and garden niche markets.

Life Cycle

Steinernematids and heterorhabditids have similar life histories. The non-feeding, developmentally arrested infective juvenile seeks out insect

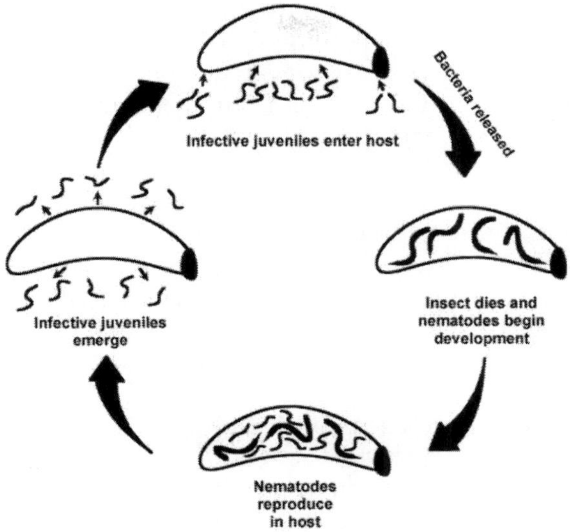

hosts and initiates infections. When a host has been located, the nematodes penetrate into the insect body cavity, usually via natural body openings (mouth, anus, spiracles) or areas of thin cuticle. Once in the body cavity, a symbiotic bacterium (*Xenorhabdus* for steinernematids, *Photorhabdus* for heterorhabditids) is released from the nematode gut, which multiplies rapidly and causes rapid insect death. The nematodes feed upon the bacteria and liquefying host, and mature into adults. Steinernematid infective juveniles may become males or females, where as heterorhabditids develop into self-fertilizing hermaphrodites although subsequent generations within a host produce males and females as well.

The life cycle is completed in a few days, and hundreds of thousands of new infective juveniles emerge in search of fresh hosts. Thus, entomopathogenic nematodes are a nematode-bacterium complex. The nematode may appear as little more than a biological syringe for its bacterial partner, yet the relationship between these organisms is one of classic mutualism. Nematode growth and reproduction depend upon conditions established in the host cadaver by the bacterium. The bacterium further contributes anti-immune proteins to assist the nematode in overcoming host defenses, and anti-microbials that suppress colonization of the cadaver by competing secondary invaders. Conversely, the bacterium lacks invasive powers and is dependent upon the nematode to locate and penetrate suitable hosts.

Production and Storage Technology

Entomopathogenic nematodes are mass produced for use as biopesticides using *in vivo* or *in vitro* methods. *In vivo* production (culture in live insect hosts) requires a low level of technology, has low startup costs, and resulting nematode quality is generally high, yet cost efficiency is low. The approach can be considered ideal for small markets. *In vivo* production may be improved through innovations in mechanization and streamlining. A novel alternative approach to *in vivo* methodology is production and application of nematodes in infected host cadavers; the cadavers (with nematodes developing inside) are distributed directly to the target site and pest suppression is subsequently achieved by the infective juveniles that emerge. *In vitro* solid culture, i.e., growing the nematodes on crumbled polyurethane foam, offers an intermediate level of technology and costs. *In vitro* liquid culture is the most cost- efficient production method but requires the largest startup capital. Liquid culture may be improved through progress in media development, nematode recovery, and bioreactor design. A variety of formulations have been developed to facilitate nematode storage and application including activated charcoal, alginate and polyacrylamide gels, baits, clay, paste, peat, polyurethane sponge, vermiculite, and water-dispersible granules. Depending on the formulation and nematode species, successful storage under refrigeration ranges from one to seven months. Optimum storage temperature for formulated nematodes varies according to species; generally, steinernematids tend to store best at 4-8 °C whereas heterorhabditids persist better at 10-15 °C.

Relative Effectiveness and Application Parameters

Growers will not adopt biological agents that do not provide efficacy comparable with standard chemical insecticides. Technological advances in nematode production, formulation, quality control, application timing and delivery, and particularly in selecting optimal target habitats and target pests, have narrowed the efficacy gap between chemical and nematode agents. Nematodes have consequently demonstrated efficacy in a number of agricultural and horticultural market segments.

Entomopathogenic nematodes are remarkably versatile in being useful against many soil and cryptic insect pests in diverse cropping systems, yet are clearly underutilized. Like other biological control agents, nematodes are constrained by being living organisms that require specific conditions to be effective. Thus, desiccation or ultraviolet light rapidly inactivates insecticidal nematodes; chemical insecticides are less constrained. Similarly, nematodes are effective within a narrower temperature range (generally between 20 °C and 30 °C) than chemicals, and are more impacted by suboptimal soil type, thatch depth, and irrigation frequency. Nematode-based insecticides may be inactivated if stored in hot vehicles, cannot be left in spray tanks for long periods, and are incompatible with several agricultural chemicals. Chemicals also have problems (e.g., mammalian toxicity, resistance, groundwater pollution, etc.) but a large knowledge base has been developed to support their use. Accelerated implementation of nematodes into IPM systems will require users to be more knowledgeable about how to use them effectively.

Therefore, based on the nematodes' biology, applications should be made in a manner that avoids direct sunlight, e.g., early morning or evening applications are often preferable. Soil in the treated area should be kept moist for at least two weeks after applications. Application to aboveground target areas is difficult due to the nematode's sensitivity to desiccation and UV radiation; however, some success against certain above-ground targets has been achieved and recently approaches have been enhanced by improved formulations. In all cases, the nematodes must be applied at a rate that is sufficient to kill the target pest; generally, 250,000 infective juveniles per m^2 of treated area is required (though in some cases an increased or slightly decreased rate may be suitable). Additionally, it is important to match the appropriate nematode species to the particular pest that is being targeted.

Appearance

Nematodes are formulated and applied as infective juveniles, the only free-living and therefore environmentally tolerant stage. Infective juveniles range from 0.4 to 1.5 mm in length and can be observed with a hand lens or microscope after separation from formulation materials. Disturbed nematodes move actively, however sedentary ambusher species (e.g. *Steinernema carpocapsae, S. scapterisci*) in water soon revert to a characteristic "J"-shaped resting position. Low temperature or oxygen levels will inhibit movement of even active cruiser species (e.g., *S. glaseri, Heterorhabditis bacteriophora*). In short, lack of movement is not always a sign of mortality; nematodes may have to be stimulated (e.g., probes, acetic acid, gentle heat) to move before assessing viability. Good quality nematodes tend to possess high lipid levels that provide a dense appearance, whereas nearly transparent nematodes are often active but possess low powers of infection.

Insects killed by most steinernematid nematodes become brown or tan, whereas insects killed by heterorhabditids become red and the tissues assume a gummy consistency. A dim luminescence given off by insects freshly killed by heterorhabditids is a foolproof diagnostic for this genus (the symbiotic bacteria provide the luminescence). Black cadavers with associated putrefaction indicate that the host was not killed by entomopathogenic species. Nematodes found within such cadavers tend to be free-living soil saprophages.

Habitat

Steinernematid and heterorhabditid nematodes are exclusively soil organisms. They are ubiquitous, having been isolated from every inhabited continent from a wide range of ecologically diverse soil habitats including cultivated fields, forests, grasslands, deserts, and even ocean beaches. When surveyed, entomopathogenic nematodes are recovered from 2% to 45% of sites sampled.

Pests Attacked

Because the symbiotic bacterium kills insects so quickly, there is no intimate host-parasite relationship as is characteristic for other insect-parasitic nematodes. Consequently, entomopathogenic nematodes are lethal to an extraordinarily broad range of insect pests in the laboratory. Field host range is considerably more restricted, with some species being quite narrow in host specificity. Nonetheless, when considered as a group of nearly 80 species, entomopathogenic nematodes are useful against a large number of insect pests. Additionally, entomopathogenic nematodes have been marketed for control of certain plant parasitic nematodes, though efficacy has been variable depending on species. A list of many of the insect pests that are commercially targeted with entomopathogenic nematodes is provided in the table below. As field research progresses and improved insect-nematode matches are made, this list is certain to expand.

Characteristics of Some Commercialized Species

Steinernema carpocapsae

This species is the most studied of all entomopathogenic nematodes. Important attributes include ease of mass production and ability to formulate in a partially desiccated state that provides several months of room-temperature shelf-life. *S. carpocapsae* is particularly effective against lepidopterous larvae, including various webworms, cutworms, armyworms, girdlers, some weevils, and wood-borers. This species is a classic sit-and-wait or "ambush" forager, standing on its tail in an upright position near the soil surface and attaching to passing hosts. Consequently, *S. carpocapsae* is especially effective when applied against highly mobile surface-adapted insects (though some below-ground insects are also controlled by this nematode). *S. carpocapsae* is also highly responsive to carbon dioxide once a host has been contacted, thus the spiracles are a key portal of host entry. It is most effective at temperatures ranging from 22 to 28°C.

Steinernema feltiae

S. feltiae is especially effective against immature dipterous insects, including mushroom flies, fungus gnats, and tipulids as well some lepidopterous larvae. This nematode is unique in maintaining infectivity at soil temperatures as low as 10°C. *S. feltiae* has an intermediate foraging strategy between the ambush and cruiser type.

Use of Nematodes as Biological Insecticides

Pest Common name	Pest Scientific name	Key Crop(s) targeted	EfficaciousNematodes *
Artichoke plume moth	*Platyptilia carduidactyla*	Artichoke	Sc
Armyworms	Lepidoptera: Noctuidae	Vegetables	Sc, Sf, Sr
Banana moth	*Opogona sachari*	Ornamentals	Hb, Sc
Banana root borer	*Cosmopolites sordidus*	Banana	Sc, Sf, Sg
Billbug	*Sphenophorus* spp. (Coleoptera: Curculionidae)	Turf	Hb,Sc
Black cutworm	*Agrotis ipsilon*	Turf, vegetables	Sc
Black vine weevil	*Otiorhynchus sulcatus*	Berries, ornamentals	Hb, Hd, Hm, Hmeg, Sc, Sg
Borers	*Synanthedon* spp. and other sesiids	Fruit trees & ornamentals	Hb, Sc, Sf
Cat flea	*Ctenocephalides felis*	Home yard, turf	Sc
Citrus root weevil	*Pachnaeus* spp. (Coleoptera: Curculionidae	Citrus, ornamentals	Sr, Hb
Codling moth	*Cydia pomonella*	Pome fruit	Sc, Sf
Corn earworm	*Helicoverpa zea*	Vegetables	Sc, Sf, Sr
Corn rootworm	*Diabrotica* spp.	Vegetables	Hb, Sc
Cranberry girdler	*Chrysoteuchia* topiaria	Cranberries	Sc
Crane fly	Diptera: Tipulidae	Turf	Sc
Diaprepes root weevil	*Diaprepes abbreviatus*	Citrus, ornamentals	Hb, Sr
Fungus gnats	Diptera: Sciaridae	Mushrooms, greenhouse	Sf, Hb
Grape root borer	*Vitacea polistiformis*	Grapes	Hz, Hb
Iris borer	*Macronoctua onusta*	Iris	Hb, Sc
Large pine weevil	*Hylobius albietis*	Forest plantings	Hd, Sc
Leafminers	*Liriomyza* spp. (Diptera: Agromyzidae)	Vegetables, ornamentals	Sc, Sf
Mole crickets	*Scapteriscus* spp.	Turf	Sc, Sr, Scap

contd....

contd....

Pest Common name	Pest Scientific name	Key Crop(s) targeted	EfficaciousNematodes *
Navel orangeworm	*Amyelois transitella*	Nut and fruit trees	Sc
Plum curculio	*Conotrachelus nenuphar*	Fruit trees	Sr
Scarab grubs**	Coleoptera: Scarabaeidae	Turf, ornamentals	Hb, Sc, Sg, Ss, Hz
Shore flies	*Scatella* spp.	Ornamentals	Sc, Sf
Strawberry root weevil	*Otiorhynchus ovatus*	Berries	Hm
Small hive beetle	*Aethina tumida*	Bee hives	Yes (Hi, Sr)
Sweetpotato weevil	*Cylas formicarius*	Sweet potato	Hb, Sc,

* At least one scientific study reported 75% suppression of these pests using the nematodes indicated in field or greenhouse experiments. Subsequent/other studies may reveal other nematodes that are virulent to these pests. Nematodes species used are abbreviated as follows: Hb=Heterorhabditis bacteriophora, Hd = H. downesi, Hi = H. indica, Hm= H. marelata, Hmeg = H. megidis, Hz = H. zealandica, Sc=Steinernema carpocapsae, Sf=S. feltiae, Sg=S. glaseri, Sk = S. kushidai, Sr=S. riobrave, Sscap=S. scapterisci, Ss = S. scarabaei.

** Efficacy of various pest species within this group varies among nematode species.

Steinernema glaseri

One of the largest entomopathogenic nematode species at twice the length but eight times the volume of *S. carpocapsae* infective juveniles, *S. glaseri* is especially effective against coleopterous larvae, particularly scarabs. This species is a cruise forager, neither nictating nor attaching well to passing hosts, but highly mobile and responsive to long-range host volatiles. Thus, this nematode is best adapted to parasitize hosts possessing low mobility and residing within the soil profile. Field trials, particularly in Japan, have shown that *S. glaseri* can provide control of several scarab species. Large size, however, reduces yield, making this species significantly more expensive to produce than other species. A tendency to occasionally "lose" its bacterial symbiote is bothersome. Moreover, the highly active and robust infective juveniles are difficult to contain within formulations that rely on partial nematode dehydration. In short, additional technological advances are needed before this nematode is likely to see substantial use.

Steinernema kushidai

Only isolated so far from Japan and only known to parasitize scarab larvae, S. kushidai has been commercialized and marketed primarily in Asia.

Steinernema riobrave

This novel and highly pathogenic species was originally isolated from the Rio Grande Valley of Texas, but has since been also been isolated in other areas, e.g., in the southwestern USA. Its effective host range runs across multiple insect orders. This versatility is likely due in part to its ability to exploit aspects of both ambusher and cruiser means of finding hosts. Trials have demonstrated its effectiveness against corn earworm, mole crickets, and plum curculio. *Steinernema riobrave* has also been highly effective in suppressing citrus root weevils (e.g., Diaprepes abbreviates and Pachnaeus species). This nematode is active across a range of temperatures; it is effective at killing insects at soil temperatures above 35°C, and can also infect at 15°C. Persistence is excellent even under semi-arid conditions, a feature no doubt enhanced by the uniquely high lipid levels found in infective juveniles. Its small size provides high yields whether using *in vivo* (up to 375,000 infective juveniles per wax moth larvae) or *in vitro* methods.

Steinernema scapterisci

The only entomopathogenic nematode to be used in a classical biological control program, *S. scapterisci* was isolated from Uruguay and first released in Florida in 1985 to suppress an introduced pest, mole crickets. The nematode become established and presently contributes to control. *Steinernema scapterisci* is highly specific to mole crickets. Its ambusher approach to finding insects is ideally suited to the turfgrass tunneling habits of its host. Commercially available since 1993, this nematode is also sold as a biological insecticide, where its excellent ability to persist and provide long-term control contributes to overall efficacy.

Heterorhabditis bacteriophora

Among the most economically important entomopathogenic nematodes, *H. bacteriophora* possesses considerable versatility, attacking lepidopterous and coleopterous insect larvae, among other insects. This cruiser species appears quite useful against root weevils, particularly black vine weevil where it has provided consistently excellent results in containerized soil. A warm temperature nematode, *H. bacteriophora* shows reduced efficacy when soil drops below 20°C.

Heterorhabditis indica

First discovered in India, this nematode is now known to be ubiquitous. *Heterorhabditis indica* is considered to be a heat tolerant nematode (infecting insects at 30°C or higher). The nematode produces high yields *in vivo* and *in vitro*, but shelf life is generally shorter than most other nematode species.

Heterorhabditis megidis

First isolated in Ohio, this nematode is commercially available and marketed especially in western Europe for control of black vine weevil and various other soil insects. *Heterorhabditis megidis* is considered to be a cold tolerant nematode because it can effectively infect insects at temperatures below 15°C.

Conservation

Conservation strategies are poorly developed and largely limited to avoiding applications onto sites where the nematodes are ill-adapted; for example, where immediate mortality is likely (e.g., exposed foliage) or where they are completely ineffective (e.g., aquatic habitats) (Lewis et al., 1998). Minimizing deleterious effects of the aboveground environment with a post-application rinse that washes infective juveniles into the soil is also a useful approach to increasing persistence and efficacy. Native populations are highly prevalent, but, other than scattered reports of epizootics, their impact on host populations is generally not well documented (Stuart et al., 2006). This is largely attributable to the cryptic nature of soil insects. Consequently, research and guidelines for conserving native entomopathogenic nematodes are in need of advancement.

Compatibility

Infective juveniles are compatible with most but not all agricultural chemicals under field conditions. Compatibility has been tested with well over 100 different chemical pesticides. Entomopathogenic nematodes are compatible (e.g., may be tank-mixed) with most chemical herbicides and fungicides as well as many insecticides (such as bacterial or fungal products). In fact, in some cases, combinations of chemical agents with nematodes results in synergistic levels of insect mortality. Some chemicals to be used with care or avoided include aldicarb, carbofuran, diazinon, dodine, methomyl, and various nematicides. However, specific interactions can vary based on the nematode and host species and application rates. Furthermore, even when a specific chemical pesticide is not deemed compatible, use of both agents (chemical and nematode) can be implemented by waiting an appropriate interval between applications (e.g., 1-2 weeks). Prior to use, compatibility and potential for tank-mixing should be based on manufacturer recommendations. Similarly, entomopathogenic nematodes are also compatible with many though not all biopesticides; interactions range from antagonism to additivity or synergy depending on the specific combination of control agents, target pest, and rates and timing of application. Nematodes are generally compatible with chemical fertilizers as well as composted manure though fresh manure can be detrimental.

Deladenus siricidicola (=Beddingia siricidicola) (Tylenchida, Neotylenchidae)

Deladenus siricidicola is a nematode parasite of the Sirex woodwasp *(Sirex noctilio,* Hymenoptera: Siricidae), an invasive pest of pine *(Pinus* spp.*)* trees. *S. noctilio,* also called the Sirex woodwasp, is a native wood boring wasp of Eurasia/North Africa, where it is not considered a pest. However, it is an invasive pest of pine trees in many other parts of the world. It was detected in New Zealand in the early 1900s, and in Tasmania and Australia by the 1950s and 1960s. By the 1980s it had moved to South America, it was in South Africa by the 1990s, and it finally was detected in North America in 2004.

In the Southern Hemisphere, around 8 million hectares of exotic pine plantations are threatened by the pest. In Australia it is estimated that, uncontrolled, the wasp could cost 16-60 million US dollars per year, and in Brazil an estimated 6.6 million US dollars per year could be lost if their control program were not in place. In the Southern US, there is an 8 billion US dollar per year pine industry that could be threatened if S. noctilio establishes in that region.

Biology

S. noctilio is a wood boring wasp which kills pine trees through the use of a symbiotic, tree-pathogenic fungus (*Amylostereum areolatum*) along with a phytotoxic mucous, which are injected into the tree along with eggs during oviposition. The mucous inhibits the tree's ability to fend off attack from the fungus, which grows throughout the wood and causes lethal wilt of the tree. As the tree is killed, the needles, particularly at the crown, droop and turn brown. A tree can be killed in as little time as a year, being rotted by the fungus. Meanwhile, eggs deposited in the wood hatch and larvae feed on the decaying wood. *S. noctilio* take 1-2 years to develop to adulthood, depending on the location of the infestation, at which point adults emerge, mate, and infest new trees.

Life Cycle

D. siricidicola has two morphologically different forms—a fungal feeding form and a parasitic form. The fungal feeding form lives in the pine tree and eats the symbiotic fungus injected into the tree by an adult *S. noctilio* female. While in this form in the tree, the nematode can reproduce for many generations in the absence of the woodwasp as it spreads throughout the tree. If, however, the nematodes detect the presence of *S. noctilio* larvae, they develop into parasitic adults. These infective adults penetrate the *S. noctilio* larvae and produce offspring, which migrate as juveniles into the reproductive organs of the host. Nematodes eventually make their way into the eggs of *S. noctilio* before the shells have hardened. *S. noctilio* are not killed by the nematode, they undergo "parasitic castration." Parasitized female *S. noctilio* emerge from trees as adults, mate, and search for suitable pine trees in which to oviposit. During oviposition, they inject fragments of the symbiotic fungus into the wood, along with phytotoxic mucous and eggs, which are at that point full of nematodes instead of woodwasps. Once the nematodes are injected into the tree by the female host, they return to the fungal feeding form. Male *S. noctilio* are considered a dead end for the nematodes, as they cannot transmit the nematodes to females during mating.

Using *D. siricidicola* for classical biological control of *S. noctilio*

To mass produce the nematode for biological control, cultures of the symbiotic *Sirex noctilio* fungus are established in flasks and inoculated with the fungal-feeding stage of the nematode. The nematodes are given time to grow and reproduce, at which point they are ready to be harvested and inoculated into trees in the field. Before nematodes are released, trap trees are made to attract *S. noctilio*. This is achieved by injecting herbicides into trees, which slowly kills them and makes them both attractive and highly susceptible to infestation by the woodwasps. The trees are injected three months before the *S. noctilio* are expected to emerge, and once infestation of the trap trees by the woodwasps is confirmed, nematodes are introduced. A special hammer is used to punch holes into the trees, and the nematodes, suspended in a gel, are injected into the holes of the tree. Infected *S. noctilio* then transfer the nematodes to subsequent trees.

These inoculations have been carried out in control programs with great success in Tasmania and Australia. It is estimated that the control program, which includes the use of the nematode, saves an estimated $16-60 million USD per year in Australia. It is also estimated that a control program including nematodes saves $6.6 million USD per year in Brazil. Success has not been universal, however, as some releases of nematodes in South Africa and South America have not lead to satisfactory control of S. noctilio. A number of possible explanations have been proposed, a few of which are climate, the presence of a defective strain of the nematode, or poor inoculation technique. Identifying what could be confounding the biological control system is an active area of research.

Mermis nigrescens Dujardin, (Phylum Nematoda, order Mermithida, family Mermithidae)

This is a large species of nematode. Adult males range from 40 to 50 mm long, while the females range from 80 to 200 mm long. The body of the worm tapers in the front, with the head slightly offset and rounded in the front. Larvae and adult males are generally cream colored, as are females if their uterus is empty. If the uterus contains eggs, the female will appear brownish or black because of the coloration of the eggs she contains. Males have two spicules, one of which is used in reproduction. Both sexes possess two laterally-placed anterior chemoreceptors called amphids, but they lack posterior phasmids. Both sexes also possess a terminal mouth with two lateral papillae close to it and four head papillae further back.

In adult females, there is an orange/red area behind the head, which is associated with a unique light sensitive organ. Mohammed et al (2007) give details of the structure and function of this organ. It serves to allow the female to know where to go when laying eggs.

On rainy mornings in May or June, large nematodes sometimes can be found up on foliage, where they are depositing their eggs. Because they are so large (up to 160 mm or about 6 inches in length), they attract attention. Under some circumstances they also function as important biological control agents, especially of grasshoppers.

Development

Embryos begin to develop in the eggs while still in the female. By the time the eggs are laid (in late spring or early summer) the offspring are already stage 2 larvae. If an egg is ingested by a grasshopper or other orthopteran, the outer layer of the egg rubs off and within minutes the larva begins to move out of the inner shell.

Once hatched, the larva pierces the gut wall of the host and enters the body cavity. Larval worms feed on host body fluids and tissues and grow. They emerge from the host in approximately 3-6 weeks. The worms develop faster in warmer temperatures. Size and species of hosts affects developmental time, and larval worms stay longer and grow larger in mature adult hosts than in nymph or young adult hosts. Female worms tend to remain in the host longer than males, and are larger at emergence than males. Worms puncture the host body wall and emerge as 4th stage larvae (the host and any remaining worms die). They crawl 15-20 cm into the soil where they molt, become mature adults and mate. Males die after mating. Females spend the winter in diapause in the soil and emerge on rainy or dew mornings in the following spring/summer to climb up on to plants and lay their eggs.

Host Range

Mermis nigrescens is normally associated with grasshoppers (Orthoptera: Acrididae, Romaleidae, Tettigoniidae) but also is reported to occur naturally in earwigs (Dermaptera), beetles (Coleoptera), caterpillars (Lepidoptera), and even honeybees (Hymenoptera). Because it is similar in appearance to other species of *Mermis*, and to *Amphimermis, Longimermis, Agamermis,* and *Hexamermis,* some host records may be inaccurate.

Geographical Range

Mermis nigrescens occurs widely in North and South America, Europe and Asia. Distribution in Africa seems to be quite limited, but uncertain. It is found in Tasmania, but not yet in Australia, though other *Mermis* spp. and related genera occur there.

Life Cycle

Mermis spp. display the curious habit of depositing their eggs on vegetation, where they are accidentally ingested by herbivorous insects; other mermithids lay their eggs in the soil and when they hatch the larvae move to the surface and penetrate the hosts. After ingestion, the *M. nigrescens* eggs quickly hatch and the immature nematodes break through the gut wall and enter the body cavity of the host insect. Juvenile *M. nigrescens* nematodes initially are about 0.75 mm long. Infection by *M. nigrescens* inhibits development of the ovaries in grasshoppers, and the host may perish before or after the nematodes emerge from the insect. After emerging, the nematodes enter the soil where they reach the adult stage after 2-4 months. They are not sexually mature for another 6 months, and may remain in the soil for up to 3 years, though a two-year life cycle is normal. *M. nigrescens* may either mate, or reproduce parthenogenetically. Males are not as long-lived as are the females. Males are about 40-60 mm long, females 60 to 160 mm. Females with fully developed eggs come to the soil surface and deposit eggs on vegetation only following protracted rainfall, often in the morning hours before the sun dries the foliage. Females produce large quantities of eggs, estimated to be up to 14,000 per female. When laden with eggs, the tan-colored females appear to have a dark stripe running the length of the body, but this dark coloration disappears as the eggs are deposited. The eggs are about 0.05 mm in diameter. They are brown or reddish brown in color. The eggs bear two clusters of branched filaments, one at each pole. As rainfall and dew evaporate, the filaments adhere to the foliage, causing the egg to remain fixed to the vegetation, where it can be consumed along with the plant. Irrigation can increase parasitism of grasshoppers by these nematodes by simulating rainfall conditions; rainfall is a prerequisite for adult female nematodes to leave the soil to deposit eggs.

Relative Effectiveness

These nematodes are most abundant where suitable insect hosts have been plentiful, usually grasshoppers. Various studies have reported high levels of parasitism in grasshoppers ranging from 20-75%. Although often considered to be absent from arid areas such as the western Great Plains of USA, they do occur where water is available. In Colorado, up to 50% of *Melanoplus femurrubrum* from irrigated cropland was found to be infected whereas the same species was unparasitized in nearby arid rangeland. Although impressive claims sometimes are made about the suppressive effects on grasshopper populations by *M. nigrescens*, their long life cycle and need for high moisture conditions reduces their reliability as a biological control agent.

Ecosystem Roles

Larva of this nematode are parasites of orthopteran insects, and can sometimes have substantial effects on host populations.

Identification

Mermis and related nematodes are often confused with horsehair worms (Nematomorpha: Gordioidea). Horsehair worms superficially resemble *Mermis nigrescens* nematodes, but may be twice as long, and usually are found near water. Horsehair worms typically are completely dark in color, with a uniform thickness throughout almost the entire length of the body. *Mermis nigrescens* nematodes are not uniformly dark and at least one end of the body tapers to a blunt point.

Phasmarhabditis hermaphrodita (Rhabditida: Rhabditidae)

Phasmarhabditis hermaphrodita is a nematode parasite of slugs which has been commercially available as a biological molluscicide in Europe since 1994. The methods of production and formulation for this nematode are broadly similar to those used for entomopathogenic nematodes.

Phasmarhabditis hermaphrodita can be detected and quantificated from soil samples by quantitative PCR. This method can also be used to distinguish it from the morphologically similar *Phasmarhabditis neopapillosa.*

Appearance

The nematodes are sold as infective juveniles (dauer larvae) approximately 1 mm long, which can be just seen with the naked eye. The infective juveniles are formulated onto clays or gels that can be added to water and applied with conventional spray equipment.

Habitat

The nematode is a soil dwelling animal and is generally found where slugs are abundant. It has been isolated from grassland and crops of wheat and oilseed rape (canola). Recent research has shown that the nematode is well adapted to life in leaf litter, compost, and organic soils.

Life Cycle

Phasmarhabditis hermaphrodita is a facultative parasite and is capable of growing on dead invertebrates, slug feces, and decaying leaf litter. The nematode can form non-feeding, developmentally arrested juveniles (dauer larvae) which are a survival stage commonly formed by rhabditid nematodes. In the case of *P. hermaphrodita,* the dauer larvae also function as the infective stage in the parasitic life cycle. Dauer larvae are third stage larvae which retain a second stage cuticle, and in which the mouth and anus are closed. In the nematode's preferred slug host, *Deroceras reticulatum,* the nematodes infect principally through a natural opening at the rear of the slug's mantle. The site of entry for other mollusc species has not been studied.

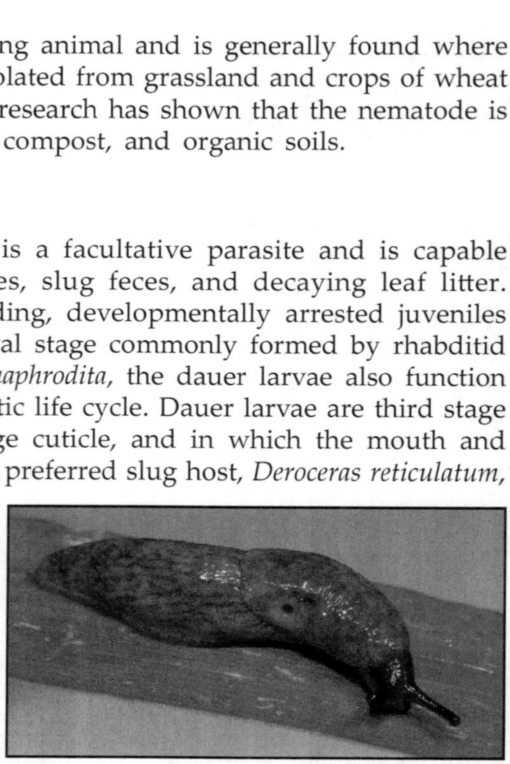

Slug pest *Deroceras reticulatum* is one of the host species for *Phasmarhabditis hermaphrodita*

Once inside the slug, the dauer larvae develop into self-fertilizing hermaphrodites which reproduce and go through another generation. During this time, certain slug species usually develop a characteristic swelling of the mantle region. The slug typically takes between 7 and 21 days to die, but its feeding is markedly reduced from the onset of infection. In laboratory experiments, a dose of nematodes roughly equivalent to half the recommended field dose inhibited slug feeding by 90% within four days of exposure. Once the slug dies, nematodes spread out over the cadaver and feed. When the food-source is depleted the nematodes again form dauer larvae which leave in search of other slugs or suitable decaying substrates.

Phasmarhabditis hermaphrodita is a bacterial-feeding nematode. Unlike entomopathogenic nematodes, no specific symbiotically associated bacterium has been isolated from this nematode to date; when reared in slug hosts, the nematode associates with complex and variable consortia of bacteria that do not influence its virulence. However, when grown *in vitro* in monoxenic culture, different bacterial partners can influence both yields and virulence of the resulting nematodes. For commercial production, *P. hermaphrodita* is always reared in monoxenic culture with the bacterium *Moraxella osloensis*.

Pests Attacked

P. hermaphrodita has been shown to infect and kill a wide variety of pest species of both slugs and snails.

Slugs

- *Arion ater* (Juveniles)
- *Arion distinctus*
- *Arion intermedius*
- *Arion lusitanicus* (Juveniles)
- *Arion silvaticus*
- *Deroceras reticulatum* (the gray field slug)
- *Deroceras laeve*
- *Deroceras panormitanum*
- *Leidyula floridana*
- *Tandonia budapestensis*
- *Tandonia sowerbyi*

SNAILS

- *Cepaea hortensis*
- *Cernuella virgata*
- *Cochlicella acuta*
- *Helix aspersa* (the brown garden snail) (Juveniles)

- *Lymnaea stagnalis*
- *Monacha cantiana*
- *Theba pisana*

The most susceptible species is the grey field slug, *Deroceras reticulatum*, which is the most widespread pest species in the world and is responsible for most agricultural and horticultural losses. While all the above species have been shown to be killed by *P. hermaphrodita* in laboratory bioassays, many of the assays used high doses and unrealistic assay conditions, so it is not clear if *P. hermaphrodita* could be used to control all the above species under field conditions. Also, laboratory bioassays indicate that body size may be an important feature of susceptibility. For example the garden snail, *Helix aspersa*, is susceptible when its body weight is less than a gram, but larger individuals are not. Similar results have been found for the large slugs *Arion ater* agg. and *A. lusitanicus*. Thus, for some large species of slug, it would be advisable to apply nematodes at the time of year when only juvenile slugs are present.

Biological control of slugs

Phasmarhabditis hermaphrodita can be used in organic farming instead of poisonous metaldehyde or carbamates.

This nematode is commercially reared in a culture with the bacterium *Moraxella osloensis* and sold as a biological molluscicide. The culture is mixed with water and applied as a drench to the surface of the soil. This is best done at a soil temperature of about 15 °C, with a minimum of 5 °C and a maximum of 20 °C. The soil should be kept damp afterwards. Young seedlings and cuttings can best be protected by applying the nematodes one week in advance of sowing or planting. The protection is expected to last for at least six weeks. Because infection of the slug reduces its feeding activities, crop protection is rapid even before host mortality occurs.

It has been shown that when released, new generations of nematodes do not continue their symbiosis with *Moraxella osloensis* but associate with a complex and variable mixture of other bacteria while still retaining their virulence

Relative Effectiveness

Phasmarhabditis hermaphrodita has been used successfully in a number of field experiments in many European countries, including the UK, Croatia, France, the Netherlands, Spain and Switzerland. There have also been successful efficacy tests in Chile and the USA. Crops in which the nematode has been tested include high value horticultural crops such as lettuce, strawberries, and glasshouse orchids and low value field crops such as Brussels sprouts, wheat, potatoes, and oilseed rape. If applied at a dose rate of 3×109/ha, the nematode gives equivalent control to chemical-standard slug pellets. The rapid feeding inhibition caused in susceptible slugs by *P. hermaphrodita* means that the nematode can be applied with the same timing as chemical molluscicides.

Non-target Effects

There is no evidence that the nematode has any undesired non-target effects. It has been tested against several earthworm and insect species and non-target mollusks. The nematode is killed at temperatures above 25°C so it is unlikely to pose a threat to mammals. The nematode can kill some aquatic snails that are considered beneficial so it is recommended that the nematode not be sprayed adjacent to ponds and waterway.

Conservation

Little is known about ways to conserve populations of *P. hermaphrodita.* Practices which typically conserve nematodes within the soil and favor build up of organic matter, e.g. reduced tillage and reduced pesticide input, are also likely to favor slugs.

Pesticide Susceptibility

Soil insecticides, nematicides, and fumigation should not be used if attempting to conserve *P. hermaphrodita.*

Lagenidium giganteum

Lagenidium giganteum is a watermold that parasitizes the larval stage of mosquitoes. This microbial parasite belongs to a group of organisms that, although they look like fungi and have a "fungal lifestyle", nonetheless are related to diatoms and brown algae. The infective stage is a motile spore that goes on a search-and-destroy mission. The basis of its host specificity is selective recognition and attachment to its mosquito hosts. If a spore encounters, for instance, a water beetle, a dragon fly larva or a rice plant, it recognizes that a susceptible host has not been encountered. It will back off from that surface and swim on looking for a suitable host.

Lagenidium giganteum (Lagenidiales) is the best-known entomopathogen belonging to the class Oomycetes. It has been described as a facultative parasite; it infects and kills available insect hosts, but also is capable of surviving saprophytically. *Lagenidium giganteum* most often infects young mosquito larvae; only occasionally does it infect older larvae and pupae. In addition, infection also has been reported to occur in certain species of gnats and biting midges. The infective propagules are asexually produced biflagellate zoospores that migrate upward and concentrate at the surface of the water to contact surface-breathing mosquito larvae. Once contact is made, the zoospores recognize and attach to the insect cuticle. Zoospores may attach over the entire larval surface or may concentrate in the head region (e.g., in the buccal cavity). After encystment and attachment, the zoospores (cysts) form delicate germ tubes that penetrate the cuticle.

Appearance

L. giganteum is most easily recognized after it has matured, and ovoid, septate cells have formed. Infected larvae are recognized by a characteristic grey-white to *almost completely white appearance.*

Habitat

Although *L. giganteum* is not an obligate parasite, and can grow vegetatively (for example, on rotting vegetation or dead insects) in the absence of its hosts, it grows much faster and is easier to isolate from mosquito larvae. It can be found in freshwater habitats supporting mosquito populations. Isolation of the parasite from indigenous larval populations has been documented primarily in the southern United States, but it has also been found in California, Cuba, Colombia, and England. It will become dormant at temperatures below ca. 16°C or above 32°C. Moderate levels of salinity or organic load prevent it from sporulating, which is necessary for mosquito infection.

Pests Attacked

The parasite will infect and kill most species of mosquito breeding in fresh water, from temperatures of 16-32°C. It will also infect the closely related dipteran *Chaoborus astictopus*, the Clear Lake gnat, and at very high concentrations, some species of daphnids. Infection of daphnids is not desireable, but this only occurs when levels of the parasite ca. 100 times greater than that recommended for operational control are reached.

Life Cycle

Infection of a larval host is initiated by motile biflagellate zoospores that selectively recognize chemical signals on the epicuticle (outer exoskeleton) of mosquitoes. After attaching (A) the zoospores inject themselves into the larva, and ramify throughout the body of the host (B). Depending upon the temperature and zoospore density, the larva dies of starvation within 1-4 days. At that time each individual cell can form an exit tube and release 10-50 asexual spores, which in turn seek out a new host (C). Alternately, two cells can fuse (D), ultimately resulting in the formation of a thick-walled dormant oospore (E). This sexual stage of *L. giganteum* can remain viable in a dehydrated state for at least 7 years. It is this spore that is responsible for multi-year recycling of the parasite even though a habitat may be dry for months or years before reflooding and colonization by mosquito larvae. Under appropriate environmental conditions, oospores will germinate, resulting in the production of infective biflagellate zoospores similar to those produced during asexual reproduction.

Relative Effectiveness

Operational levels of mosquito control have been obtained by ground or aerial application of *L. giganteum* at rates ranging from ca. 0.9 × 1010 to 5 × 1010 CFU's (colony forming units) per hectare. The application rate depends upon the susceptibility and developmental rate of the target species, and habitat characteristics (temperature, organic load and salinity). For instance, control of floodwater *Aedes* species in early fall, in which there is synchronous hatch of large numbers of eggs in relatively cold water, would require treatment at the higher rates. At the other extreme, for 3- to 4-month control of very susceptible mosquito species such as *Culex tarsalis* breeding in rice fields at very low densities, the lower rates can be used.

Zoospores do not have a cell wall, so are much too fragile to be applied directly in a breeding habitat; therefore, either presporangia, oospores, or a mixture of both are applied. Sporulation then occurs in the treated site to initiate infection.

The sexual stage of *L. giganteum* has many advantages in an operational control program including multi-year stability, resistance to desiccation and abrasion, and it's inherent slow-release characteristics. Unfortunately, fermentation yields of oospores remain 2 orders of magnitude below that of the less stable mycelial (asexual, presporangial) stage. Research is continuing on improving oospore yields, which would be much more useful in large scale operational mosquito control programs.

L. giganteum is a facultative parasite, and has been grown in large fermentation tanks using inexpensive culture media. Multi-hectare aerial applications have been made in rice fields and in refuges for migratory birds, primarily in California, where both efficacy and safety to nontarget organisms has been documented. Because the mycelial cells are relatively rugged, the low volume spraying equipment used by many mosquito abatement agencies for application of insecticides is readily adapted to application of this parasite.

Pesticide Susceptibility

Except for the dormant oospore stage, the parasite is very susceptible to organophosphates, carbamates, herbicides, fungicides, and other chemicals commonly used either by mosquito abatement districts, or applied to crops (such as rice and soybeans) that are associated with mosquito breeding. IC50 values for inhibition of mycelial growth commonly range from ca. 50-5,000 ppm. Fortunately, recommended application rates for many of these chemicals are below levels that are toxic to the parasite. Organic solvents such as toluene and oil commonly used as diluents of the active ingredients of pesticides are also toxic to *L. giganteum*.

5

Antagonists of Plant Disease and Food Spoilage Microorganisms

Ampelomyces quisqualis (Deuteromycetes)

The fungus Ampelomyces quisqualis is a naturally occurring hyperparasite of powdery mildews. It infects and forms pycnidia (fruiting bodies) within powdery mildew hyphae, conidiophores (specialized spore-producing hyphae), and cleistothecia (the closed fruiting bodies of powdery mildews). This parasitism reduces growth and may eventually kill the mildew colony. *A. quisqualis* has been the subject of numerous investigations on biological control of powdery mildews for over 50 years.

Appearance

Parasitized mildew colonies are dull, flattened, and off-white to gray in color. Spore production of the powdery mildew is reduced or absent in parasitized areas of the colony. Dead leaf tissue may be adjacent to parasitized areas of the mildew colony, as the outer leaf tissues die in response to death of the contained powdery mildew cells.

The pycnidia of *A. quisqualis* vary in shape depending upon the organ in which they form. Within conidiophores they are pear-shaped, within hyphae they are spindle-shaped, and within cleistothecia they may be nearly spherical.

Uninfected hyphae and conidiophores of powdery mildews are transparent, but turn translucent-white soon after infection. Once the mycoparasite has begun to produce pycnidia, the hyphae and conidiophores swell to several times their normal diameter, and the amber color of the pycnidial wall of the mycoparasite may be noticed through the cell walls of the host.

Parasitized cleistothecia are typically dull, fawn colored, flaccid, and range in size from 64 to 130 micrometers in diameter. Pycnidia contain cylindrical to spindle-shaped, occasionally curved, and two-spotted conidia that are 7.5-9.0 X 2.5-3.5 micrometers. The conidia may be exuded when parasitized tissues are exposed to free water or humidity near saturation.

Habitat (Crops)

A. quisqualis exists in colonies wherever its many hosts are distributed.

Pests Attacked

The mycoparasite is specific to powdery mildews (Erysiphales), but has an extremely broad host range within this diverse group of important plant pathogens. It has been recorded on more than 64 species in the genera *Brasilomyces, Erysiphe, Leveillula, Microsphaera, Phyllactinia, Podosphaera, Sphaerotheca,* and *Uncinula,* as well as the anamorphic genera *Oï dium* and *Oï diopsis*. These reports represent powdery mildews which attack 256 plant species within 172 genera in 59 families, and which occur in 28 countries around the world.

Life Cycle

The pathogen overwinters or survives intercrop periods as pycnidia. The relative importance of pycnidia within different host structures (i.e., hyphae vs conidiophores vs cleistothecia) as survival structures is unknown. When it rains, conidia are exuded from the ripe pycnidia and are dispersed by rain-splash to mildew colonies. Free water is required for infection, and infection is favored by warm temperatures (20-30° C). Under favorable conditions, infection can occur in less than 24 hours.

The mycoparasite directly penetrates the walls of hyphae, conidiophores, and immature cleistothecia, but may be unable to infect mature cleistothecia. For approximately 7-10 days, the mycoparasite spreads within the hyphae of the mildew colony without killing it. Thereafter, the process of pycnidial formation begins, and is then completed within 2-4 days. Infected cells generally die soon after pycnidial formation begins. Secondary cycles of infection result from conidia released during subsequent rain events.

Several generations may occur during a single growing season.

Relative effectiveness

Efficacy in controlling powdery mildews has been quite variable. The fact that some powdery mildew must be present to serve as a base of establishment for the mycoparasite makes this a less than ideal means to control powdery mildews on crops with no tolerance for disease (either extremely high susceptibility, high value, no tolerance for defects, or all of the foregoing). However, a number of examples of acceptable disease control have been reported for greenhouse and field-grown vegetable crops. Repeated applications of the mycoparasite are generally necessary, and high humidity and rainfall aid in spread to developing mildew colonies.

Pesticide susceptibility

Isolates differ in their susceptibility to fungicides. However, isolates tolerant to many of the most commonly used fungicides have been collected and are available for use. Susceptibility to other pesticides is poorly understood.

Conservation

A. quisqualis may survive in parasitized cleistothecia on the bark of deciduous perennial hosts, as well as on colonies on fallen leaves and crop debris. However, reintroduction on an annual basis is likely to be necessary for commercially acceptable disease control in annual or greenhouse crops.

Phlebiopsis

(=Phanerochaete, =Peniophora, =Phlebia, =Peniophora) gigantea
(Basidiomycetes: Corticiaceae)

Phlebiopsis gigantea is a common saprophytic fungus that causes white rot of conifer logs and stumps. It is used as a biological control of annosum root rot, caused by *Heterobasidion* spp., in western Europe. In the United States, *P. gigantea* was used as a biological control for annosum root rot from the 1970s to 1995. Currently, the Environmental Protection Agency is reviewing the status of *P. gigantea* as a pesticide.

Annosum root rot is primarily a problem in conifer plantations that have been partially cut (e.g. thinned). It is a pathogenic white-rot fungus that colonizes freshly cut stumps by air-borne spores. The fungus then progress from the stump down the stump's root system and infects adjacent trees through root graphs or root contacts. A tree can be infected by *Heterobasidion* and show no aboveground symptoms until at least half the root system or the root collar is infected. Above-ground symptoms include declining tree (often in groups), windthrown trees, and resin-soaking or white-string rot at the root collar. Bark beetles often occurs in infected trees. Below-ground symptoms of annosum root rot progress from resin-soaking to white-stringy rot of tree roots. The fruiting body of *Heterobasidion* spp. forms irregular masses of white fungus between bark scales that develop into a leathery conk. The conks are brown with a white margin and a cream colored lower surface. These fruiting bodies usually form on the base of the tree in the duff layer.

Appearance

The fruiting body of *P. gigantea* grows flat (1mm thick or less) along the wood surface. The surface has minute hairs (pilose) and appears pinkish-buff or marble-gray in color.

Habitat

P.gigantea occurs throughout North America, Central America, Europe, East Africa, and southern Asia. The fungus plays an important role in the decomposition of conifer debris.

Host Range

P. gigantea is found on the wood and bark of gymnosperms (notably on species of *Pinus, Picea, Abies,* and *Tsuga*). There have been reports of *P. gigantea* causing sap rot of both gymnosperm and angiosperm pulpwood in storage and of wood products.

The annual fruiting body of *P. gigantea* produces sexual spores (basidiospores) that become airborne and spread to woody debris via the wind and rain. *P. gigantea* also has an asexual stage that produces spores (oidia) from mycelial fragmentation. The asexual spores are easily produced on artificial media in the laboratory. However, their production and role in nature is unclear.

Hyphal Interference

P. gigantea has an interesting mode of action as a biological control agent. The hyphae of this fungus antagonize the hyphae of *Heterobasidion* spp.(and some other fungi) on contact - a phenomenon termed hyphal interference. Any hypha of *Heterobasidion* hypha that makes contact with a hypha of *P. gigantea* shows rapid, localized disruption: the protoplasm becomes disorganized and its membrane integrity is affected. This is shown in the figure on the right, where the dye neutral red was added to an agar plate: several annosum hyphae have grown up to a hypha of *P. gigantea* and have locally taken up the dye whereas their normal membrane integrity would exclude this dye.

Relative effectiveness

Artificial inoculation of pine stumps with *P. gigantea* has been shown to significantly decrease the presence of annosum root rot in several studies in England (Rishbeth 1963, Tubby et al. 2008) and in the southern United States (Hodges 1964, Ross and Hodges 1981). *P. gigantea* is also effective in colonizing spruce stumps. Although *P. gigantea* is common in the environment, natural levels of inoculum were found to be too low and sporadic to effectively control *Heterobasidion* spp. without artificial inoculation.

Pseudomonas syringae (saprophytic strain) and "Fruit Yeasts"

The antagonist bacterium, *Pseudomonas syringae,* is a saprophytic strain which was originally isolated from an apple leaf and can be frequently isolated from apple fruit. "Fruit yeasts" consist of many pink and white yeasts belonging to *Deuteromycota, Basidiomycota* and *Ascomycota,* isolated from a variety of fruits including apple, pear, and citrus. Both *P. syringae* and fruit yeasts grow exponentially on wounded apple and pear tissue and prevent necrotrophic pathogens such as *Penicillium expansum* or *Botrytis cinerea* from entering fruit through wounds and initiating infection and subsequent decay.

Habitat

P. syringae is a nutritionally versatile organism. There are 41 pathovars of *P. syringae* which are able to cause diseases on various plants. The antagonistic strain is their non-pathogenic counterpart and is antagonistic to pathogens on many plants. It can grow well on wounded plant tissue and can control a variety of diseases on different fruits, including pome fruits, banana, and citrus fruits as well as vegetables. The fruit's surface is a natural habitat for many yeasts which dominate fruit microflora, especially close to harvest. The yeasts grow rapidly on this substrate, which has a high concentration of readily available carbon sources, such as those contained in juices leaking from wounded fruits. Mummified fruit and soil are important over-wintering sites for yeasts and serve as an inoculum source for developing fruit. The composition of the yeast microflora on the fruit changes as the fruit matures. Humans have been taking advantage of yeasts naturally occurring on fruit from ancient times by allowing them to ferment fruit to make wine or cider. Although yeasts are common inhabitants on many fruits, most of our knowledge concerns yeasts on grape and apple because interest in vinification and cider making has stimulated research in these areas.

Diseases controlled

The discovery of the potential of *P. syringae* and yeasts that naturally occur on apple and pear as biocontrol agents against fungi that cause postharvest decay of the fruits was followed by the isolation of microorganisms from various fruits and the testing of their antagonistic potential in many systems in laboratories world wide. This resulted in the discovery of many bacterial and yeast antagonists which are effective against postharvest diseases on a variety of fruits. Some of these appear to have commercial potential. In developing this page, we are focusing initially on antagonists developed in our program. These include *P. syringae*, which has been commercialized, and "fruit yeasts" in general, because many of them are effective in controlling postharvest decays of apple and pear.

P. syringae can control blue mold caused by *P. expansum*, gray mold caused by *B. cinerea*, and Mucor rot caused by *Mucor* spp. on apple and pear. It can also control blue mold caused by *Penicillium italicum*, and green mold caused by *P. digitatum* on citrus fruit. Another strain (ESC-10) of this bacterium, superior in controlling decays of citrus fruit, has been isolated by EcoScience, Corp. Strain ESC-11 has been shown to reduce crown rot of banana, which is caused by a complex of fungi, including *Fusarium semitectum* and *F. moniliforme*, and it also reduced Fusarium dry rot on potato caused by *F. sambucinum*. This strain also prevented growth of the foodborne pathogen, *Escherichia coli*, in apple wounds. This pathogen can grow quickly on damaged apple tissue and consumption of unpasteurized apple cider contaminated with this bacterium has caused outbreaks of illnesses in recent years.

"Fruit yeasts" can control postharvest decays on many different fruits. We have isolated a yeast, *Sporobolomyces roseus*, and many other pink and white yeasts (most of them identified by isolate numbers) that can control blue mold and gray mold of pome fruits. In addition, other investigators have reported a number of "fruit yeasts" that were effective antagonists against postharvest fruit decay. A white yeast, *Candida oleophila*, has been commercialized for the control of blue mold and green mold on citrus fruit and blue mold and gray mold on apple. In general, many "fruit yeasts" can reduce fruit decays to varying degrees but only a few can reduce decay below the threshold acceptable to the fruit industry and at concentrations realistic for commercial development.

Relative effectiveness

Postharvest biological control of blue mold and gray mold on Golden Delicious apples.

Biological control in the postharvest environment has significant advantages over that under field conditions because the two most important factors effecting biocontrol, temperature and relative humidity are constant and under strict control. In addition, the targeted area (fruit) is easily accessible. These factors greatly reduce the variability of biocontrol and also make this system more amenable to manipulation. The effectiveness of biocontrol may be influenced by postharvest treatments, such as diphenylamine and ethoxyquin, which are used on pome fruits against superficial scald, a physiological disorder, and additives, such as flotation salts and waxes. In addition, fruit maturity and cultivar may effect biocontrol. However, if these factors are considered in developing a biocontrol strategy, highly consistent and commercially acceptable control of fruit decay developing from infections at wound sites can be achieved. Postharvest biocontrol can also be integrated with other non-fungicidal methods such as calcium infiltration or heat treatment, which by themselves cannot provide adequate control but in combination with biocontrol may increase the performance margin of the latter .

Trichoderma spp., including *T. harzianum, T. viride, T. koningii, T. hamatum* and other spp. Deuteromycetes, Moniliales (asexual classification system)

(*Ascomycetes, Hypocreales,* usually *Hypocrea* spp., are sexual anamorphs, this life stage is lacking or unknown for biocontrol strains)

Trichoderma species are important biological control agents used in plant disease management. They are imperfect fungi, with their teleomorph perfect stage belonging to Hypocreales of Ascomycetes. They are capable of secreting hydrolytic enzymes and cause mycoparasitism on fungal pathogens of plants. Besides, they produce antibiotics and toxins that inhibit the fungal plant pathogens. They are better rhizosphere colonizers than the plant pathogens, hence competing with other organisms for food and space in rhizosphere there by reducing the chances of colonization by pathogenic fungi.

The study of the influence of environmental parameters on *Trichoderma* strains is of great importance when planning the application of strains for biocontrol. Strains used in applied biological control should have better stress tolerance levels than the targeted plant pathogens causing plant diseases. The important stress factors include high temperature, fungicides, salinity, low moisture and soil pH. Though there are few reports on the susceptibility of *Trichoderma* species to these stresses *in vitro,* there is lack of comprehensive information on the mechanism(s) of stress tolerance in *Trichoderma* to these stresses, especially at biochemical and molecular level.

In India, there is great diversity in soil characteristics especially with respect to soil pH. *Trichoderma* strains are able to grow in a wide range of pH from 2.0 to 6.0 with maximal growth rates at 4.0, the optimum range being 4.6 to 6.8. However there is a need to have strains specifically for saline soils and sodic soils. Similarly, in major parts of country, high soil temperature is an important factor for the survival of *Trichoderma* species. The residual toxicity due to fungicides used for the control of soil borne pathogens is an important environmental concern. Therefore the improvement of stress tolerance in *Trichoderma* strains could result in increasing their efficacy against plant pathogenic fungi even under unfavourable environmental conditions. For this purpose the understanding of molecular mechanisms with respect to stress tolerance is essential.

Trichoderma spp. are fungi that are present in nearly all soils and other diverse habitats. In soil, they frequently are the most prevalent culturable fungi.

They are favored by the presence of high levels of plant roots, which they colonize readily. Some strains are highly rhizosphere competent, i.e., able to colonize and grow on roots as they develop. The most strongly rhizosphere competent strains can be added to soil or seeds by any method. Once they come into contact with roots, they colonize the root surface or cortex, depending on the strain. Thus, if added as a seed treatment, the best strains will colonize root surfaces even when roots a meter or more below the soil surface and they can persist at useful numbers up to 18 months after application. However, most strains lack this ability.

In addition to colonizing roots, *Trichoderma* spp. attack, parasitize and otherwise gain nutrition from other fungi. Since *Trichoderma* spp. grow and proliferate best when there are abundant healthy roots, they have evolved numerous mechanisms for both attack of other fungi and for enhancing plant and root growth. Several new general methods for both biocontrol and for causing enhancement of plant growth have recently been demonstrated and it is now clear that there must be hundreds of separate genes and gene products involved in these processes. A recent list of mechanisms follows.

- Mycoparasitism
- Antibiosis
- Competition for nutrients or space
- Tolerance to stress through enhanced root and plant development
- Solubilization and sequestration of inorganic nutrients
- Induced resistance
- Inactivation of the pathogen's enzymes

Taxonomy and genetics

Most *Trichoderma* strains have no sexual stage but instead produce only asexual spores. However, for a few strains the sexual stage is known, but not among strains that have usually been considered for biocontrol purposes. The sexual stage, when found, is within the Ascomycetes in the genus *Hypocrea*. Traditional taxonomy was based upon differences in morphology, primarily of the asexual sporulation apparatus, but more molecular approaches are now being used. Consequently, the taxa recently have gone from nine to at least thirty-three species.

Most strains are highly adapted to an asexual life cycle. In the absence of meiosis, chromosome plasticity is the norm, and different strains have different numbers and sizes of chromosomes. Most cells have numerous nuclei, with some vegetative cells possessing more than 100. Various asexual genetic factors, such as parasexual recombination, mutation and other processes contribute to variation between nuclei in a single organism (thallus). Thus, the fungi are highly adaptable and evolve rapidly. There is great diversity in the genotype and phenotype of wild strains.

While wild strains are highly adaptable and may be heterokaryotic (contain nuclei of dissimilar genotype within a single organism) (and hence highly variable), strains used for biocontrol in commercial agriculture are, or should be, homokaryotic (nuclei are all genetically similar or identical). This, coupled with tight control of variation through genetic drift, allows these commercial strains to be genetically distinct and nonvariable. This is an extremely important quality control item for any company wishing to commercialize these organisms.

Pathogens controlled

So far as the author is aware, different strains of *Trichoderma* control every pathogenic fungus for which control has been sought. However, most *Trichoderma* strains are more efficient for control of some pathogens than others, and may be largely ineffective against some fungi. The recent discovery in several labs that some strains induce plants to "turn on" their native defense mechanisms offers the likelihood that these strains also will control pathogens other than fungi.

Life cycle

Fungal thalli are shown in the figure at the beginning of this web page. The organism grows and ramifies as typical fungal hyphae, 5 to 10 μm in diameter. Asexual sporulation occurs as single-celled, usually green, conidia (typically 3 to 5 μm in diameter) that are released in large numbers. Intercalary resting chlamydospores are also formed, these also are single celled, although two or more chlamydospores may be fused together.

Pesticide susceptibility

Trichoderma spp. possess innate resistance to most agricultural chemicals, including fungicides, although individual strains differ in their resistance. Some lines have been selected or modified to be resistant to specific agricultural chemicals. Most manufacturers of *Trichoderma* strains for biological control have extensive lists of susceptibilities or resistance to a range of pesticides.

Uses of *Trichoderma*

These versatile fungi are used commercially in a variety of ways, including the following:

Foods and textiles

Trichoderma spp. are highly efficient producers of many extracellular enzymes. They are used commercially for production of cellulases and other enzymes that degrade complex polysaccharides. They are frequently used in the food and textile industries for these purposes. For example, cellulases from these fungi are used in "biostoning" of denim fabrics to give rise to the soft, whitened fabric—stone-washed denim. The enzymes are also used in poultry feed to increase the digestibility of hemicelluloses from barley or other crops.

Biocontrol agents

As noted, these fungi are used, with or without legal registration, for the control of plant diseases. There are several reputable companies that manufacture government registered products.

Plant growth promotion

For many years, the ability of these fungi to increase the rate of plant growth and development, including, especially, their ability to cause the production of more robust roots has been known. The mechanisms for these abilities are only just now becoming known.

Some of these abilities are likely to be quite profound. Recently, we have found that one strain increases the numbers of even deep roots (at as much as a meter below the soil surface). These deep roots cause crops, such as corn, and ornamental plants, such as turfgrass, to become more resistant to drought.

Perhaps even more importantly, our recent research indicates that corn whose roots are colonized by *Trichoderma* strain T-22 require about 40% less nitrogen fertilizer than corn whose roots lack the fungus. Since nitrogen fertilizer use is likely to be curtailed by federal mandate to minimize damage to estuaries and other oceanic environment (there are a number of other sites on the web dealing with this topic, search for sites dealing with the 'dead zone') the use of this organism may provide a method for farmers to retain high agricultural productivity while still meeting new regulations likely to be imposed.

As a source of transgenes

Biocontrol microbes, almost by definition, must contain a large number of genes that encode products that permit biocontrol to occur. Several genes have been cloned from *Trichoderma* spp. that offer great promise as transgenes to produce crops that are resistant to plant diseases. No such genes are yet commercially available, but a number are in development.

These genes, which are contained in *Trichoderma* spp. and many other beneficial microbes, are the basis for much of "natural" organic crop protection and production.

6

Predators

The arthropod predators of insects and mites include beetles, true bugs, lacewings, flies, midges, spiders, wasps, and predatory mites. Insect predators can be found throughout plants, including the parts below ground, as well as in nearby shrubs and trees. Some predators are specialized in their choice of prey, others are generalists. Some are extremely useful natural enemies of insect pests. Unfortunately, some prey on other beneficial insects as well as pests.

Insect predators can be found in almost all agricultural and natural habitats. Each group may have a different life cycle and habits. Although the life history of some common predators is well studied, information on the biology and relative importance of many predatory species is lacking. In this document, we have included the more common and better understood beneficial predators.

Major characteristics of arthropod predators

- adults and immatures are often generalists rather than specialists
- they generally are larger than their prey
- they kill or consume many prey
- males, females, immatures, and adults may be predatory
- they attack immature and adult prey

Relative Effectiveness

Most beneficial predators will consume many pest insects during their development, but some predators are more effective at controlling pests than others. Some species may play an important role in the suppression of some pests. Others may provide good late season control, but appear too late to suppress the early season pest population. Many beneficial species may have only a minor impact by themselves but contribute to overall pest mortality. Often too, the role of the beneficial predators has not been adequately studied.

Surveys of agricultural systems give an indication of the potential number and diversity of predators in a crop. For example, over 600 species of predators in 45 families of insects and 23 families of spiders and mites have been recorded in Arkansas cotton. Eighteen species of predatory insects (not including spiders and mites) have been found in potatoes in the northeastern United States. There may be thousands of predators per acre, in addition to many parasitoids. Although the impact of any one species of natural enemy may be minor, the combined impact of predators, parasitoids, and insect pathogens can be considerable.

BEETLES

Lady Beetles (Coleoptera: Coccinellidae)

Lady beetles, ladybugs, or ladybird beetles are the Coccinellidae, a family of beetles. All species are protected by noxious fluids based on cyanide, and most have warning colouration, such as red with black spots.

They are often called *lady bugs* or *ladybirds*, but biologists prefer the term *lady beetle.*

Coccinellids are found worldwide, with over 5,000 species. Most species are carnivorous, feeding mainly on the *true bugs*, the Hemiptera. Their favourite food is pests such as aphids (greenfly) or scale insects. Their larvae are also voracious (greedy) eaters of greenfly.

Harmonia axyridis (or the Harlequin ladybug) was introduced into North America from Asia in 1988 to control aphids but is now the most common species as it is out-competing many of the native species.

Lady beetles are among the most visible and best known beneficial predatory insects. Over 450 species are found in North America. Some are native and some have been introduced from other countries.

Most lady beetles in North America are beneficial as both adults and larvae, feeding primarily on aphids. They also feed on mites, small insects, and insect eggs. The two exceptions are the introduced Mexican bean beetle, *Epilachna varivestis*, and the squash beetle, *Epilachna borealis*. The adults and larvae of both species feed on plants.

Coccinella septempunctata

Appearance

Lady beetles are usually red or orange with black markings. Some lady beetles are black, often with red markings. They have alligator-like larvae.

Habitat (Crops)

Many crops benefit from lady beetles. They are helpful for growers of vegetables, grain crops, legumes, strawberries, and tree crops; however any crop that is attacked by aphids will benefit from these beetles.

Pests Attacked

Most lady beetles found on crops and in gardens are aphid predators. Some species prefer only certain aphid species while others will attack many aphid species on a variety of crops. Some prefer mite or scale species. If aphids are scarce, lady beetle adults and larvae may feed on the eggs of moths and beetles, and mites, thrips, and other small insects, as well as pollen and nectar. They may also be cannibalistic. Because of their ability to survive on other prey when aphids are in short supply, lady beetles are particularly valuable natural enemies.

Appearance

Adult lady beetles are small, round to oval, and dome-shaped. The most well known have black markings on red, orange, or yellow forewings, but some are black. The area immediately behind the head, the pronotum, may also have a distinctive pattern. The color and pattern of markings for each species may vary, but can aid identification.

Life Cycle

Lady beetles overwinter as adults, often in aggregations along hedgerows, beneath leaf litter, under rocks and bark, and in other protected places including buildings. In spring, the adults disperse in search of prey and suitable egg laying sites. This dispersal trait, especially strong in migratory species such as the commercially available convergent lady beetle, can affect the reliability of control by released adult beetles.

Female lady beetles may lay from 20 to more than 1,000 eggs over a one to three month period, commencing in spring or early summer. Eggs are usually deposited near prey such as aphids, often in small clusters in protected sites on leaves and stems. The eggs of many lady beetle species are small (about 1 mm; 1/25"), cream, yellow, or orange, and spindle-shaped. They resemble those of Mexican bean beetle and Colorado potato beetle, but are usually smaller.

Lady beetle larvae are dark and alligator-like with three pairs of prominent legs. Depending on the species and availability of prey, larvae grow from less than 1 mm (1/25") to about 1 cm (3/8") in length, typically through four larval instars, over a 20 to 30 day period. Large larvae may travel up to 12 m (about 40') in search of prey. The larvae of many species are gray or black with yellow or orange bands or spots.

The last larval instar remains relatively inactive before attaching itself by the abdomen to a leaf or other surface to pupate. Pupae may be dark or yellow-orange. The pupal stage may last from three to 12 days depending on the temperature and species. The adults emerge, mate, and search for prey or prepare for hibernation, depending on the availability of prey and time of year. Adults may live for a few months to over a year. The more common species typically have one to two generations per year.

Relative Effectiveness

Lady beetles are voracious feeders and may be numerous where prey are plentiful and broad-spectrum insecticide use is limited. Lady beetles need to eat many aphids per day so that they can lay eggs. The convergent lady beetle may eat its weight in aphids every day as a larva and consume as many as 50 aphids per day as an adult. Sevenspotted lady beetle adults may consume several hundred aphids per day and each larva eats 200 to 300 aphids as it grows. Once the adults and larvae have eliminated an aphid colony, they will search for additional food.

Lady beetles are effective predators if aphids are abundant (high pest density) but are thought to be less effective at low pest densities. There may also be some crop damage before lady beetles have an impact on an aphid population.

Most lady beetle species are active from late spring to early fall if food is available. The adults of some species, such as *Coleomegilla,* may be attacked by a parasitoid wasp. Affected beetles are lethargic and may appear to be "attached" to a surface by loose silken threads.

Pesticide Susceptibility

Lady beetles may be tolerant of some chemical insecticides applied at recommended field rates. But tolerance may be influenced by prior exposure of the natural enemy population to the chemical, and there is considerable variation. For example, *Coleomegilla maculata* has been shown to be susceptible to carbaryl and methamidophos at doses that might be used to control aphids in potatoes. Overwintering adults may be less susceptible to chemical insecticides than active adults and larvae.

Conservation

Lady beetles are mobile as adults and reasonably so as larvae, and they are generalists. They will not remain on a plant, or in the vicinity, once the readily accessible prey has been consumed.

Some species, in particular *Coleomegilla maculata,* consume pollen as an important part of their adult diet. A source of nectar and pollen, or an artificial substitute, in the vicinity of the crop may attract adult beetles and may reduce dispersal of this and other lady beetle species. Lady beetle adults also benefit from high humidity and nearby shelter, for protection from adverse weather and to provide overwintering sites.

Early season populations of some lady beetles may develop on aphid-infested perennials or shrubs. Collection and redistribution of lady beetles is effective but time consuming. The beetles should be handled gently and placed in groups at the base of plants, rather than broadcast. Hibernating adults, congregating in protected spaces, should not be disturbed as they are vulnerable to attack by predators and parasitoids if uncovered.

Chilocorus kuwanae (Coleoptera: Coccinellidae)

Chilocous kuwanae is a small black lady beetle imported from Korea as part of a biological control program for the euonymus scale, *Unaspis euonymi.* This scale is a serious pest that kills wintercreeper euonymus, a popular North American ground cover. *C. kuwanae* has been released by the USDA in most states east of the Mississippi River, five states in the central U.S., and in California. Field surveys show that *C. kuwanae* is widespread from Alabama to southern New England and westward to the Appalachian mountains.

Appearance

C. kuwanae is a black lady beetle, about 3 mm long, with red spots. It looks quite similar to the native *Chilocorus stigma,* but it can be distinguished by the color, shape and location of the spot on its wings. Spots of *C. kuwwanae* tend to be deep red and rectangular and located near the center of the wing. In contrast, spots of *C. stigma* tend to be more orange-yellow, round and oriented more toward the head of the beetle.

Habitat

Vegetation that is attacked by scales.

Pests attacked

In North America., euonymus scales and other scales, such as the San Jose scale. In China, Korea and Japan, where it is common, *C. kuwanae* helps to keep several species of armored scales under control in citrus groves and on landscape shrubs.

Life Cycle

Beetles spend the winter as adults in leaf litter at the base of scale-infested plants. They become active and feed on scales when temperatures exceed 50°F in the spring. Adults lay bright orange eggs, singly or in small groups under the scale covers. Eggs hatch into brown larvae covered with black spines. Larvae will flip over or chew through scale covers to feed on the fleshy scale body. After 3 larval stages, larvae move to the underside of a leaf or to a crack or crevice on a twig or branch. There they become immobile and the larval skins split to form pupae. In the laboratory, it takes about a month to mature from egg to adult. In the midwest during the growing season, there are usually 3 generations.

Relative Effectiveness

C. kuwanaeis noted for its voracious appetite. Each larval *C. kuwanae* must feed on over 100 scales to become an adult. In the United States, this appetite has caused the beetle to rapidly deplete the euonymus scale population at landscape release sites. Like many other lady beetles, *C. kuwanae* just flies away when food gets scarce. However, other scales such as San Jose scale and Oystershell scale can be used by adults as a food source.

Conservation

In areas where this beetle is established, you can help it along by leaving leaf litter on the ground so that beetles have a safe place to spend the harsh midwestern winters. Despite reductions in numbers due to the exceptionally cold winter of 1993-94, some beetles survived where temperatures dipped to -27°F. In addition, you should adopt standard practices that conserve natural enemies. In the landscape, this means limiting pesticide use and using materials with short residual toxicities, such as horticultural oils or insecticidal soaps when pesiticides must be applied. Flowering plants may also contribute to the residence time of *C. kuwanae* adults. They have been reported to feed on nectar and pollen in the landscape.

Conservation of natural enemies is arguably the most important concept in the practice of biological control and fortunately is also one of the easiest to understand. Simply put, conservation of natural enemies means avoiding practices which are harmful to natural enemies and implementing practices which benefit them. Sounds like good common sense! The tricky part comes in understanding exactly what practices are harmful and how beneficial practices can be integrated into your production system. That requires that we understand the biology of the natural enemies and are willing to modify our practices to accommodate them.

Chilocorus stigma (Say) (Coleoptera: Coccinellidae)

Twice-stabbed ladybeetle

Nearly all species in the genus *Chilocorus* are predaceous on scale insects, although some will accept aphids or adelgids as prey. Most species are tropical, but seven species of *Chilocorus* are native to the United States. Two species, *C. kuwanae* and *C. bipustulatus*, have been introduced for biological control. However, the native *Chilocorus* stigma is the only one that occurs in most of the United States. It does not occur west of the Sierra Nevada. *C. stigma* is synonymous with *C. bivulnerus* Mulsant.

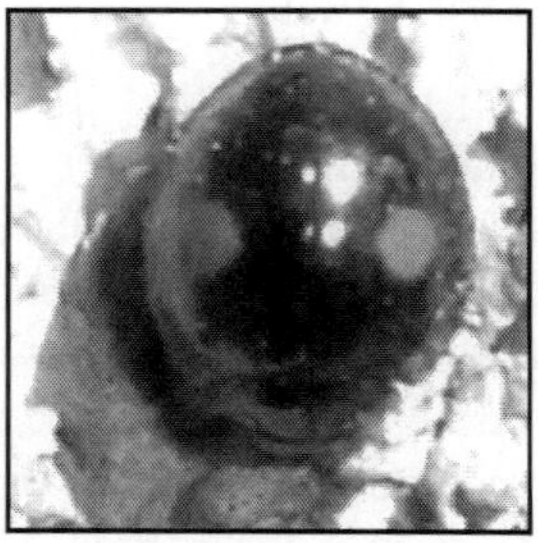

Chilocorus stigma

Appearance

C. stigma adults appear shiny black with a large red spot in the center of each elytron. Adults average 3.75-5.0 mm in length. The body is completely black except for the abdomen, which is yellow or red. A few other native *Chilocorus* species closely resemble *C. stigma*, but for the most part they are found in California where *C. stigma* does not occur. The introduced species *C. kuwanae* appears similar but can be distinguished by the appearance of the spot on its wings (see page on *C. kuwanae*). Larvae are black or grey and spiny in appearance. Eggs are small (about 1.1 mm long), orange, and laid on their sides either singly or in small groups.

Habitat

Chilocorus species typically prefer arboreal habitats. They have been identified as beneficial natural enemies in orchards, tree plantations, and forests. In Michigan, *C. stigma* is often found in Christmas tree plantations, and forests where scale insect infestations are found. It is a beneficial insect, and is useful in both natural wood stands and commercial forests such as orchards and citrus groves. It is beneficial against non native species. An introduced Hemlock pest, the elongate hemlock scale (*Fiorinia externa* Ferris), which has been doing considerable damage to Hemlock trees throughout North America, can be moderately controlled by the presence of *C. stigma*. *C. stigma* is currently not a lady beetle that can be sold for commercial use in orchards or on farms.

Pests Attacked

The genus *Chilocorus* consists mostly of armored scale insect predators. *C. stigma* is an omnivorous predator of several scale insects, aphids, and mealybugs. It has been reported often as a predator of the pine needle scale (*Chionaspis pinifoliae*) (Cumming 1953, Neilsen and Johnson 1973, DeBoo and Weidhaas 1976). It has been observed attacking beech bark scale (*Cryptococcus fagisuga*) in Michigan. Muma (1955) found it associated with the Florida red scale (*Chrysomphalus aonidum*) in citrus groves.

Life Cycle

C. stigma completes two generations in the northern U.S. (DeBoo and Weidhaas 1976). In Florida, it can complete several generations a year (Muma 1955). In cold areas, overwintering adults become active in early spring (April or May). Mating begins shortly and continues for about three weeks. Eggs are laid soon afterwards. Larvae emerge in late May and undergo four instars before pupating (Muma 1955). DeBoo and Weidhaas (1976) observed that in New York State, the overwintering adults continue to feed through June and early July, after first generation larvae have hatched. First generation adults mature in early to mid July. Mating and oviposition continue as before, and second generation larvae are observed beginning in mid July through early August. Adults emerge in midsummer and overwinter in ground litter.

Relative Effectiveness

Some workers report that *C. stigma* can be a highly effective predator of pine needle scale. Researcher states that *C. stigma* greatly reduced populations of pine needle scale in localized areas of Saskatchewan. Others dispute the claim that it is important in the absence of other predators. Researchers found it only rarely associated with the pine needle scale in north central Colorado. Its impact on the beech bark scale in Michigan is not yet known.

Pesticide Susceptibility

Like most beneficial insects, *C. stigma* is susceptible to broad-spectrum insecticides. Using alternative control treatments such as dormant or horticultural oils may be less harmful to the lady beetles.

Conservation

Avoiding or reducing the use of broad-spectrum insecticides will help conserve beneficial natural enemies such as *C. stigma*. Preserving refuges such as windbreaks or forested edges will give *C. stigma* a place to survive and recolonize scale-infested fields or orchards after disturbances. In Christmas tree plantations, selective spraying of the most severely infested individual trees near market, when possible, is one way to conserve natural enemies.

Coccinella septempunctata (Coleoptera: Coccinellidae)

Sevenspotted Lady Beetle

The sevenspotted lady beetle, repeatedly introduced to North America from Europe for the biological control of aphids, was established in the early 1970s in New Jersey, apparently from an accidental introduction. It has since spread naturally or been introduced to many northeastern and north central states. *C. septempunctata* may be a more effective predator than some native lady beetle species, displacing them in some areas.

Appearance

Comparatively large (7-8 mm) with a white or pale spot on either side of the head. The body is oval, and has a domed shape. The spot pattern is usually 1-4-2, black on the orange or red forewings. Lady beetle larvae are dark and alligator-like with three pairs of prominent legs, growing to 7-8 mm in length. Eggs are spindle shaped and small, about 1 mm long.

Habitat (Crops)

Aphid infested crops, including potatoes, legumes, sweet corn, alfalfa, wheat, sorghum, and pecans.

Pests Attacked

Reported prey include pea, cowpea, green peach, potato, corn leaf, melon aphids, and greenbug.

Life Cycle

Adults overwinter in protected sites near the fields where they fed and reproduced. In spring, emerging beetles feed on aphids before laying eggs. Females may lay from 200 to more than 1,000 eggs over a one to three month period commencing in spring or early summer. Eggs are usually deposited near prey such as aphids, often in small clusters in protected sites on leaves and stems. The eggs are small (about 1 mm) and spindle-shaped.

C. septempunctata larvae grow from about 1 mm to 4-7 mm in length over a 10 to 30 day period depending on the supply of aphids. Large larvae may travel up to 12 m in search of prey. A second generation may appear about a month later. The pupal stage may last from three to 12 days depending on the temperature.

In the northeastern United States, there are one to two generations per year before the adults enter winter hibernation. Development from egg to adult may take only two to three weeks, and adults, most abundant in mid- to late summer, live for weeks or months, depending on the location, availability of prey, and time of year.

Conservation

C. septempunctata is spreading to new areas each season. Conservation can best be accomplished by following integrated pest management guidelines as outlined in the tutorial of this guide.

Pesticide Susceptibility

Tolerance to some pesticides at recommended application rates is likely. Overwintering adults may be less susceptible than active adults and larvae.

Coleomegilla maculata (Coleoptera: Coccinellidae)

This lady beetle, a North American native, is found throughout most of New York, southern Ontario, and southern New England and across the south to the mid-western states. It can be very common and is sometimes called the pink spotted lady beetle.

Appearance

C. maculata, is a medium-sized (about 5-6 mm), pink to red, oval beetle with six spots on each forewing. The area behind the head is often pink or yellowish with two large triangular black marks. *C. maculata* larvae are dark and alligator-like with three pairs of prominent legs, growing to 5-6 mm in length. Eggs are spindle shaped and small, about 1 mm long.

Habitat (Crops)

Wheat, sorghum, alfalfa, soybeans, cotton, potatoes, sweet corn (tassels may supply both pollen and small prey), peas, beans, cole crops, tomatoes, asparagus, apples and other crops attacked by aphids and other reported prey.

Pests Attacked

C. maculata adults and larvae are important aphid predators but also prey on mites, insect eggs, and small larvae. Unlike most lady beetles, plant pollen may constitute up to 50% of the diet. Reported prey include pea, green peach, melon (cotton), cabbage, and potato aphids and greenbug; eggs of European corn borer, imported cabbageworm, fall webworm, and corn earworm; asparagus beetle, Mexican bean beetle, and Colorado potato beetle eggs and larvae. In trials to assess this lady beetle for control of Colorado potato beetle, it appeared to prefer aphids over beetle eggs and larvae.

Life Cycle

Adults overwinter in large aggregations beneath leaf litter and stones along hedgerows or in protected sites along crop borders, especially those of fields planted to corn in the previous season. They emerge from early to mid-spring and disperse, often by walking along the ground, to seek prey and egg laying sites in nearby crops.

Female lady beetles may lay from 200 to more than 1,000 eggs over a one to three month period commencing in spring or early summer. Eggs are usually deposited near prey such as aphids, often in small clusters in protected sites on leaves and stems. Larvae grow from about 1 mm to 5-6 mm in length and may wander up to 12 m in search of prey. The larva attaches itself by the abdomen to a leaf or other surface to pupate. The pupal stage may last from 3 to 12 days depending on the temperature.

C. maculata adults may be found from April to late September and can be the most commonly observed lady beetles in corn, potatoes, and mixed crops. They may be especially abundant towards the end of the season when the adults aggregate in preparation for mating and winter hibernation. There are from two to five generations per year.

Conservation

Because pollen is an essential component of the diet of *Coleomegilla,* the planting or preservation of refuges, or interplantings, of early-flowering species with a high pollen load may be beneficial especially to provide a food source during late spring before the build up of aphids. Flowering dandelions, for example, have been recorded as a heavily used pollen source for dispersing adults in late spring potato fields.

Cryptolaemus montrouzieri (Coleoptera: Coccinellidae)

Mealybug Destroyer

This beetle was imported into the United States in 1891 from Australia by one of the early biological control pioneers, Albert Koebele, to control citrus mealybug in California. Although *C. montrouzieri* initially devastated the citrus mealybug populations in citrus groves, it was unable to survive the winter except in coastal areas.

Appearance

Cryptolaemus montrouzieri is a small (about 3-4 mm long), dark brown lady beetle with a tan to orange head and posterior. The larvae grow up to 1.3 cm in length and have woolly appendages of wax (their true legs are barely visible underneath) which makes them resemble mealybugs, although they are about twice as large as the adult female citrus mealybug. *C. montrouzieri* eggs are yellow.

Habitat

Citrus groves in the coastal areas of California, interiorscapes, and greenhouses. In addition, *C. montrouzieri* is released seasonally into inland citrus orchards.

Pests Attacked (Host Range)

C. montrouzieri attacks citrus and closely related mealybugs and some soft scales, including hemispherical scale and its relatives. It is considered an important predator of citrus and long-tailed mealybug in greenhouses and interior plantscapes and is being introduced in a biocontrol program in the West Indies to control pink hibiscus mealybug.

Life Cycle

Adult females lay eggs among the cottony egg sack of adult female mealybugs. Eggs hatch into larvae in about 5 days at 27°C (80°F). The three larval stages last from 12-17 days during which the larvae feed on mealybug eggs, young crawlers, and the honeydew produced by mealybugs.

C. montrouzieri pupates in sheltered places on stems or on greenhouse structures. Adults emerge after 7-10 days and live four months. Four days after emerging, adult females begin to lay eggs, and the lifetime total may be as many as 400. Adults may live up to 2 months.

Relative Effectiveness

C. montrouzieri is a voracious feeder of mealybug in both the larval and adult stages - a single larva may consume up to 250 small mealybugs. They are most effective when mealybug populations are high, and repeated releases may be necessary if mealybug populations are low. They require cottony egg masses for egg-laying (long-tailed mealybugs do not have cottony egg masses). Although adults and young larvae prefer to feed on mealybug eggs, older larvae will attack any mealybug stage. Adults can fly and cover large areas to search for food. If mealybugs are scarce, they will fly off in search of other related insects, e.g. aphids and soft scales, although reproduction is substantially greater on mealybugs.

Because *C. montrouzieri* cannot survive cold winters, they must be reintroduced into orchards where mealybugs were a problem the previous year in the early spring. The exception is moderate coastal regions.

Like other ladybeetles, *C. montrouzieri* tends to disperse when released. In indoor sites, keep windows and vents closed the day of release. Recent studies have shown that adults and larvae will spend more time searching a leaf for mealybugs if it has honeydew than if honeydew is absent.

Although mealybug destroyers will not persist after the mealybug population has been controlled in small greenhouse areas, in more complex interior areas where there are multiple infestation sites and alternate prey, they do perist. In Indianapolis, a single release of 100 beetles has persisted for over 5 years. If birds exist in interiorscapes, however, they will feed on adult beetles.

Conservation

It is important to remember that *C. montrouzieri* superficially resembles its prey, the mealybug. This can cause problems if large numbers of larvae are present on crops that are ready to be shipped or in indoor situations (such as malls) where the public may be unaware of the benefits of natural enemies.

Harmonia axyridis (Coleoptera: Coccinellidae)

Multicolored Asian Lady Beetle

The multicolored Asian lady beetle is an introduced biological control agent that is spreading rapidly throughout the United States. Despite the fact it is an effective biological control agent in agricultural and landscape environments, it has become a major nuisance to homeowners because of its habit of invading houses and buildings in large numbers while searching for protected sites to overwinter in the fall (mid-October to early November) and appearing again on warm, sunny days in February and March. In addition, beetles may get in picnic food and drinks, "swarm" like bees and land on people. For helpful insights in dealing with the Asian lady beetle if it becomes a nuisance, please see the information compiled from Ohio State University Despite these annoying traits, H. axyridis preys upon many species of injurious soft-bodied insects such as aphids, scales, and psyllids and is thus considered beneficial to growers and agriculturalists.

The USDA made several releases throughout the eastern United States in an attempt to introduce this beetle into North America. It took a long time for releases made in Louisiana and Mississippi in 1979 and 1980 to become established and spread, but by 1994, *H. axyridis* was found in Alabama, Georgia, South Carolina, and Florida. They are now widely disseminated in North America with frequent reports from the south, northeast, midwest, and as far as Oregon.

Although "multicolored Asian lady beetle" is the common name officially accepted by the Entomological Society of America, several other common names are also found in the literature: Halloween lady beetle (because of its pumpkin orange color and large populations often observed around Halloween), Japanese lady beetle (because Japan was the country of origin for specimens released in the southeastern United States), and Asian lady beetle.

Harmonia axyridis is a large coccinellid beetle. Its colour ranges from yellow-orange to black, and the number of spots between none and 22. It is native to eastern Asia, but has been introduced to North America and Europe to control aphids and scale insects. It is now common, well known, and spreading in those regions, and has also established in South Africa and widely across South America.

It is commonly known as the harlequin ladybird (because it occurs in numerous colour forms). It is also known in North America as the multicolored Asian lady beetle, and (because it invades homes in October in preparation for overwintering) as Halloween lady beetle. In Japan, it is not generally distinguished from the seven-spot ladybird which is also common there.

When the species first arrived in the UK, it was labelled in jest as "the many-named ladybird", because among the names listed were: multivariate, southern, Japanese, and pumpkin ladybird.

Appearance

H. axyridis occurs in many color forms. Adults are strongly oval and convex, about 6 mm long, and 5 mm wide. North American populations have a mix of individuals ranging in color from pale yellow-orange to bright red-orange, with or without black spots on the wing covers. The head, antennae, and mouthparts are generally straw-yellow but are sometimes tinged with black. The pronotum is similarly straw-yellow with up to 5 black spots or with lateral spots usually joined to form 2 curved lines, an M-shaped mark, or a solid trapezoid. The wing covers are generally yellow-orange in unspotted beetles. In fully spotted beetles, each wing cover has 10 black spots.

Eggs are bright yellow, laid in clusters of about 20 on the undersides of leaves. Larvae are elongate, somewhat flattened, and adorned with strong tubercles and spines. The mature larva (or fourth instar) is strikingly colored: the overall color is mostly black to dark bluish-gray, with a prominent bright yellow-orange patch on the sides of abdominal segments 1 to 5.

Habitat (Crops)

A variety of nursery, ornamental, and field crops in North America, including cotoneaster, rose, Christmas trees, apple, pecan, alfalfa, wheat, cotton, tobacco, and small grains.

Pests Attacked

Many species of injurious soft-bodied insects such as aphids, scales, and psyllids, including pecan aphids, red pine scale, balsam twig aphids, and pine bark adelgid.

Life Cycle

It is believed that females overwinter in protected sites unmated, with the majority of the population mating later in the spring. Eggs generally hatch in 3 to 5 days. The larval stage lasts 12 to 14 days, and the pupal stage, which takes place on leaves, lasts 5 to 6 days. In cool spring weather, development from egg to adult can take 36 days or longer. After emergence, adults can live as long as 2 to 3 years under optimal conditions.

Relative Effectiveness

In Japan, *Harmonia axyridis* is considered primarily an arboreal species and is common on various aphid-infested trees and bushes such as maple, walnut, willow, and rose; it is also an important predator of various destructive scales in Japan and mainland China. An adult is capable of consuming 90 to 270 aphids per day, and each larva can consume 600 to 1,200 aphids during its development. *H. axyridis* is a promising biological control agent of several insect pests on a wide variety of ornamnental and agricultural crops. Its large, and even explosive, populations are probably caused by the massive abundance of prey (predominantly aphids and scales), apparent lack of competition from native lady beetles, and apparent lack of native natural enemies. Scientists predict that multicolored Asian lady beetle populations will become more balanced when its prey numbers decrease and *Harmonia* itself falls prey to native natural enemies.

Pesticide Susceptibility

Pesticides should not be used for controlling populations of *H. axyridis*. The best technique for managing lady beetles is first to prevent their entry into houses and other buildings by sealing cracks and openings around windows, doors, siding, and utility pipes with a quality silicone or silicone-latex caulk. Similarly, repair door and window screens or other openings to the outdoors using regular window screening (about 18 × 16 mesh size). If beetles still gain entry into living spaces, they should be removed using a broom and dustpan, or vacuum cleaner, and released outdoors. When using a broom and dustpan, gently collect the beetles to avoid alarming them. If alarmed, they may discharge a yellow fluid that can stain walls, paint, and fabrics, and that has an unpleasant odor. All lady beetles do this reflex bleeding when alarmed. The blood comes out of the leg joints.

Biology and Behaviour

H. axyridis becomes dormant in cooler months, though it will move around whenever the temperature reaches about 10 °C (50 °F). Because the beetles will use crevices and other cool, dry, confined spaces to overwinter, significant numbers may congregate inside walls if given a large enough opening.

These beetles make some use of pheromones to "call" each other, allowing for the large gatherings often seen in the autumn. This is exploited by the makers of *H. axyridis* traps. However, many cues are visual, both at long (picking out light-coloured structures that are distinct from their surroundings) and short (picking out pre-existing aggregations to join) ranges, while nonvolatile long-chain hydrocarbons laid down by previous aggregations also play a significant role in site selection. Both are more important than volatile pheromones.

They often congregate in sunlit areas because of the heat available, so even on fairly cold winter days, some of the hibernating beetles will "wake up" because of solar heating. These large populations can be problematic because they can form swarms and linger in an area for a long time. These beetles can form groups that tend to stay in upper corners of windows. This beetle has been also found to be attracted to dark screening material for its warmth. This beetle has good eyesight, and will come back from where it was removed, and is known to produce a small bite if provoked.

H. axyridis, like other lady beetles or ladybirds, uses isopropyl methoxy pyrazine as a defensive chemical to deter predation, but also contains this

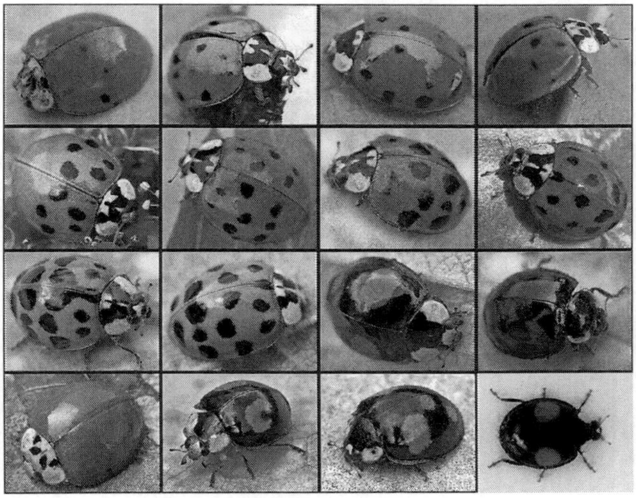

Different patterns

chemical in its hemolymph at much higher concentrations than many other such species, along with species/genus-specific defensive compounds such as harmonine. These insects will "reflex bleed" when agitated, releasing hemolymph from their legs. The liquid has a foul odour (similar to that of dead leaves) and can cause stains. Some people have allergic reactions, including allergic rhinoconjunctivitis when exposed to these beetles. Sometimes, the beetles will bite humans, presumably in an attempt to acquire salt, although many people feel a pricking sensation as a lady beetle walks across the skin, which is just the pressure from the ladybird's feet. Bites normally do no more harm than cause irritation, although a small number of people are allergic to bites.

Control

Numerous methods of control have been investigated in areas where this beetle has been introduced and causes a threat to native species and biodiversity and to the grape industry. Methods of control include insecticides, trapping, removal of aggregates of beetles, and mechanically preventing entry to buildings. Methods under development involve the investigation of natural parasites and pathogens, including the use of parasitic sexually transmitted mites and fungal diseases.

H. axyridis traps are available that contain the pheromones used by the beetles to attract each other into large gatherings. The best methods for dealing with them in private homes involve sealing openings they may enter. Sweeping and vacuuming are considered effective methods for removing them from homes, though this should be done carefully so as not to trigger reflex bleeding. A nylon stocking placed inside the vacuum cleaner's hose, secured with a rubber band, allows the beetles to be "bagged" rather than collected inside the machine. A trap designed for indoor use was developed which attracts the beetles with a light and seals them in a removable bag, though as the beetles are not strongly attracted to light, this does not work particularly well.

Hippodamia convergens (Coleoptera: Coccinellidae)

Convergent Lady Beetle

This is one of the best known and most common American lady beetles and is found from southern Canada to South America. This is the species sold by insectaries for aphid control.

Appearance

Convergent lady beetle adults are slightly elongated in shape and can range from 4-7 mm in length. They have a prominent black and white pattern behind the head, and black spots on red forewings. Beetles may have a full complement of 13 spots or they may have only a few. The white lines that converge behind the head are common to all individuals.

Habitat (Crops)

Wheat, sorghum, alfalfa, vegetables, greenhouse crops, orchard crops, and most crops attacked by aphids.

Pests Attacked

Adults and larvae prey mainly on aphids. Reported prey include cotton, pea, melon, cabbage, potato, green peach, and corn leaf aphids. If aphids are scarce, beetles and larvae may feed on small insect larvae, insect eggs, mites and, occasionally, nectar, and honeydew secreted by aphids and other sucking insects. Convergent lady beetles have been recorded as predators of asparagus beetle eggs and larvae and potato psyllids.

Life Cycle

These beetles can adjust their life cycle according to the availability of aphids. Resident populations in the eastern United States are active throughout spring and summer if aphids are present.

Females may lay from 200 to more than 1,000 eggs over a one to three month period commencing in spring or early summer. Eggs are usually deposited near prey such as aphids, often in small clusters in protected sites on leaves and stems. The eggs are small (about 1 mm) and spindle-shaped.

Lady beetle larvae are dark and alligator-like with three pairs of prominent legs. They grow from about 1 mm to 4-7 mm in length over a 10 to 30 day period. Large larvae may travel up to 12 m in search of prey. The pupal stage may last from three to 12 days depending on the temperature.

One to two generations occur each year, depending on the length of the season, before the adults enter winter hibernation, usually in a protected site. Development from egg to adult may take only two to three weeks, and adults live for weeks or months, depending on the location, availability of prey, and time of year.

In the western United States, adult convergent lady beetles typically spend up to nine months, from May to February, hibernating in large aggregations in mountain valleys, far from potential food sources. In spring, the adults disperse in search of prey and suitable egg laying sites. This dispersal trait is especially strong in *H. convergens*.

Pesticide Susceptibility

Tolerance to some pesticides at recommended application rates is likely. Overwintering adults may be less susceptible than active adults and larvae.

Conservation

Whenever possible, conserve local populations by minimizing the use of broad-spectrum insecticides, providing refuges for shelter and overwintering.

The conservation of natural enemies is probably the most important and readily available biological control practice available to growers. Natural enemies occur in all production systems, from the backyard garden to the commercial field. They are adapted to the local environment and to the target pest, and their conservation is generally simple and cost-effective. With relatively little effort the activity of these natural enemies can be observed. Lacewings, lady beetles, hover fly larvae, and parasitized aphid mummies are almost always present in aphid colonies. Fungus-infected adult flies are often common following periods of high humidity. These natural controls are important and need to be conserved and considered when making pest management decisions. In many instances the importance of natural enemies has not been adequately studied or does not become apparent until insecticide use is stopped or reduced. Often the best we can do is to recognize that these factors are present and minimize negative impacts on them. If an insecticide is needed, every effort should be made to use a selective material in a selective manner.

Sasajiscymnus (formerly *Pseudoscymnus) tsugae* (Coleoptera: Coccinellidae)

Among the most widespread and effective predators of hemlock woolly adelgid (HWA), *Adelges tsugae* (Annand), in its native homeland of Japan is a previously non-described coccinellid. In 1992, Dr. Mark McClure collected this ladybird beetle from adelgid-infested hemlocks in 13 of 37 forests and at 11 of 37 ornamental sites in 12 prefectures throughout Honshu, Japan. Dr. Hiroyuki Sasaji and Dr. McClure described this new beetle in 1997 and named it *Pseudoscymnus tsugae*. In 2004, the genus was renamed *Sasajiscymnus* by Vandenberg as a replacement for a name already in use for a genus of shark. Henceforth, this adelgid predator is known as *Sasajiscymnus tsugae*.

A federal permit for the release of *S. tsugae* for biological control of the hemlock woolly adelgid in Connecticut was issued in April 1995. The first release of *S. tsugae* occurred in a town park in Windsor, Connecticut, in the spring of 1995. From 1995 to 2007, more than 176,000 adult beetles were reared and released in adelgid-infested hemlock forests throughout Connecticut. Since 1997, after the transfer of a starter colony to New Jersey, approximately 2 million beetles have been reared and released throughout the range of hemlock woolly adelgid in eastern North America from Georgia to Maine. This adelgid predatorhas now been released in 16 states, with the majority of releases occurring in the southern Appalachians in the last 8 years.

Appearance

The oval-shaped egg of *S. tsugae* is about 0.48 mm long by 0.25 mm wide and reddish-orange in color within an opalescent sheath or chorion. Eggs are often laid singly or in small groups in cracks and crevices in the bark and in bud scales. Larvae change from reddish-brown to gray and grow from about 1.1 mm to 2.7 mm during their four instars of development. The pupa is reddish-brown and is about 1.9 mm long and 1.1 mm wide. The newly emerged adult is light golden brown before darkening to jet black within a day. Adults are oval-shaped and are about 1.7 mm long by 1.1 mm wide. The body is entirely black and is pubescent on its dorsal surface.

In Japan, adults and larvae were found on branches of *Tsuga diversifolia* and *T. sieboldii* infested with *A. tsugae* throughout Honshu; adults were collected from sweep net samples of a grassy meadow in Fukui Prefecture. In the US, all life stages have been recovered from *T. canadensis* at release sites in Connecticut and Virginia. In Connecticut, adults could be found on infested hemlock foliage from April to November and even January during the warmest winters. Larvae were found in June and July through September in Connecticut.

Pests Attacked (Host Range)

Sasajiscymnus tsugae is only known to attack *A. tsugae* in nature. However, laboratory experiments revealed that *S. tsugae* also can feed and develop on other adelgid species, including balsam woolly adelgid, *A. piceae;* Cooley spruce gall adelgid, *A. cooleyi;* and pine bark adelgid, *Pineus strobi. S. tsugae* will also oviposit on balsam woolly adelgid and pine bark adelgid and can adapt to survive and reproduce on balsam woolly adelgid in the field.

Life Cycle

The life cycle of *S. tsugae* is well synchronized with that of its prey. Both insects have two generations each year in the field. Adult beetles overwinter, and spring egg laying by the adults normally coincides with peak egg laying and hatching of overwintering adelgids. Furthermore, a second generation of beetles occurs in June around the time that the second generation of adelgids appears. Also, when adelgids are inactive for about 14 weeks during the summer, adult ladybugs are able to survive by feeding on dormant young adelgids. Three or more generations of *S. tsugae* can be reared each year in the laboratory under controlled temperature conditions with the provision of developing adelgids. *S. tsugae* larvae and adults will feed on all stages of hemlock woolly adelgid available. *S. tsugae* is the only introduced predator that is bivoltine and able to survive the summer on dormant settled first instars of HWA.

Relative Effectiveness

Adults and larvae of *S. tsugae* are highly mobile and voraciously feed on all life stages of *A. tsugae.* During its development, each beetle larva consumes about 500 adelgid eggs or from 50 to 100 adelgid nymphs, depending upon the size of prey attacked. Adults can live for more than one year and may consume about 50 adelgid nymphs each week during times of peak feeding activity. Each female beetle lays nearly 300 eggs in her lifetime.

Sasajiscymnus tsugae is an important predator of *A. tsugae* in Japan, killing 86 to 99% of the adelgids at the 24 sites were it was studied. Experiments conducted at four sites in Connecticut and one site in Virginia from 1995 through 1997 revealed that releasing relatively few adult beetles (2,400-3,600) into an infested hemlock forest reduced adelgid densities by 47-88% on release trees in only 5 months. In the field, *S. tsugae* mated, reproduced, and dispersed from release trees and became established in the surrounding hemlock forest. Field recoveries in Connecticut showed that *S. tsugae* overwintered successfully from 1996-2005 under a wide variety of climatic conditions. Adults and larvae were recovered at 65% of Connecticut release sites one to six years after the initial release. Severe winters in the northeast in 2003, 2004, and 2009 have impacted survival of hemlock woolly adelgid, keeping Connecticut forest populations of this pest patchy and low and generally non-damaging. While *S. tsugae* has been difficult to recover at ground level using lower crown beat sampling, bucket truck sampling has shown that adults and larvae could be found between 5-20m in the upper canopy in Connecticut and New Jersey, indicating that adults are highly mobile and disperse upwards. Environmental conditions of cool, wet summers and cold winters have aided hemlock recovery in Connecticut. Recovery in adelgid-infested stands has been recorded statewide in Connecticut since 2005, while hemlock mortality has been minimal since 2002. More recently, in 2008 and 2009, adults and larvae of *S. tsugae* were recovered during lower canopy beat sampling in 22% of release sites in the Great Smoky Mountains National Park. Recoveries were made in 21% of the sites sampled, with most beetles recovered from older release sites. These studies suggest that *S. tsugae* releases have been successful and that the beetle has excellent potential for further biological control of *A. tsugae.*

Conservation

The conservation of natural enemies is probably the most important and readily available biological control practice available to growers. Natural enemies occur in all production systems, from the backyard garden to the commercial field. They are adapted to the local environment and to the target pest, and their conservation is generally simple and cost-effective. With

relatively little effort the activity of these natural enemies can be observed. Lacewings, lady beetles, hover fly larvae, and parasitized aphid mummies are almost always present in aphid colonies. Fungus-infected adult flies are often common following periods of high humidity. These natural controls are important and need to be conserved and considered when making pest management decisions. In many instances the importance of natural enemies has not been adequately studied or does not become apparent until insecticide use is stopped or reduced. Often the best we can do is to recognize that these factors are present and minimize negative impacts on them. If an insecticide is needed, every effort should be made to use a selective material in a selective manner.

Rodolia cardinalis (Coleoptera: Coccinellidae)

VEDALIA BEETLE

Over the winter of 1888-1889 a lady beetle called vedalia beetle was introduced into California from Australia to combat cottony cushion scale, *Icerya purchasi* - also an introduced species. Cottony cushion scale was causing infestations so severe in California citrus groves that growers were pulling out their trees and burning them. Orchard values were plummeting. Yet, by the fall of 1889, the pest was completely controlled in the areas of introduction. The vedalia beetle literally saved the California citrus industry, and since the California success, it has been exported to many other parts of the world, often with equally successful results.

Rodolia cardinalis

The introduction of the vedalia beetle is considered to be the beginning of classical biological control. The interest of scientists, growers and the public in this project was due to its spectacular success which was striking because the financial threat to the California citrus industry was acute; the pest itself was showy and its damage was obvious and critical; growers took the initiative and applied the natural enemies themselves; the natural enemies were visibly voracious and active; and the destruction of the pest and the recovery of the trees was evident within months. The cost of the project was about $1500.

The adult has a semispherical body, 2-4mm long, covered with dense, short hairs. It is reddish-purple with black spots localized in several parts of its body, forming a net of contours between the spots. The head, posterior part of the prothorax across the full width, and the scutellum are all black.

There are typically five black spots on the elytron. Four of those are arranged on the dorso-lateral part of the elytron. The two anterior spots form an roughly half-moon shaped oval with the convexity directed towards the suture of the elytron. The two posterior ones make a more irregular shape, formed by the intersection of two circular spots. Finally, the fifth spot covers the length of the elytron's suture, englarging towards the posterior stretch.

The antenna are short and slightly clubbed, composed of 8 items, of which the proximal is markedly pulled aside. The legs have an extended and irregularly flattened tibia, forming a space housing the tarsus when at rest. The tarsus is composed of 3 tarsomeri.

The larva is around 5mm long, coloured the same red as the mature beetle, with black spots on the thorax. The left side has a series of tubercles, each with short bristles on. The pupa is 4-5mm long. It is a red which darkens with age in as the abdomen darkens.

Appearance

Adults are very small, densely pubescent, red and black lady beetles, about 2.5-4 mm (1/16-3/16 inch) long. Eggs are red, young larvae are reddish, and pupal cases are whitish.

Habitat

Acacia, boxwood, citrus, magnolia, Nandina, olive, Pittosporum, and rose - all species attacked by cottony cushion scale. The vedalia beetle is established in California and Florida and may also be found in disjunct locations of the southern U.S.

Pests Attacked

R. cardinalis is specific to cottony cushion scale. Adults and mature larvae feed on all scale stages; young feed on eggs.

Life Cycle

Adult beetles lay eggs underneath the scale or attached to scale egg sacs. There are 8 generations per year in the cooler coastal areas of California, and 12 generations per year in the hot, dry inland areas. A female will lay from 150-190 eggs during her lifetime.

Relative Effectiveness

Together with an imported parasitoid, *Cryptochetum iceryae, Rodolia cardinalis* keeps California populations of cottony cushion scale at extremely low levels in orchards and on ornamentals.

The original 514 beetles imported over the winter of 1888-89 multiplied so quickly that by June 12, officials were able to distribute10,555 vedalia beetles to 208 growers around the state. The rapidity and extent of the subsequent control was reported to be nearly unbelievable. One grower who had abandoned hope for his young orange trees was able to harvest two to three boxes of oranges from each tree at the end of the season in 1889. By October, it had become very difficult to find a living specimen of *Icerya* in the vicinity of Los Angeles, the area where the introductions were made.

Many diverse climates have become home to beetles descended from those introduced to California. From the original 514 individuals, thousands of beetles have been sent to other locations around the world. In one instance, 3 individuals from Italy, also descendants of the California beetles, provided the genetic basis for a population established in France. (This population was later augmented with more beetles from Italy, California, and Portugal.) The California population is the sole basis for colonies established in Egypt, Cyprus, the Soviet Union, Portugal, Puerto Rico, Venezuela, Peru, Chile, Hawaii, the Philippines, Guam, Uruguay, Argentina, Taiwan, and Palau, suggesting this population has an enormous ability to adapt.

Pesticide Susceptibility

The vedalia beetle is very sensitive to Baythroid, an insecticide being used for citrus thrips control. Baythroid kills the beetles and prevents them from laying eggs for about one month. In contrast, some other insecticides used for citrus thrips control, Veratran, Agri-Mek, and Success do not seem to have any noticeable effect on vedalia beetles. The vedalia beetle is also very sensitive to the new insect growth regulators, Esteem (=Knack, pyriproxifen) and Applaud (buprofezin), which are applied for California red scale control. The insect growth regulators prevent vedalia from pupating and emerging as adults and in the case of Esteem they prevent the adults from laying fertile eggs.

Orchards adjacent to orchards sprayed with growth regulators had severe cottony cushion scale problems in 1998 and 1999, indicating the beetles are extremely sensitive to these pesticides either because of drift or because the beetles are moving in and out of treated orchards. The effects of the insecticides are long lasting (4-6 months). After sprays were applied in the summer of 1998, live vedalia beetles could not be found in the San Joaquin Valley of California for 9 months. After fewer sprays were applied in the summer of 1999, live vedalia were found in the fall of 1999 in many orchards. Researchers at the University of California Kearney Agricultural Center are conducting experiments to learn how to rear vedalia beetles so that they can be released in problem orchards and are continuing to study in detail how the insect growth regulators affect vedalia.

Over the years, there have been other times that vedalia beetle has been eliminated from orchards by insecticide applications. During the years that DDT was used extensively, the vedalia beetle population in orchards in the Central Valley of California was decimated. It took three years for growers to modify their spray programs and recolonize their orchards with the beetle. In the meantime, some trees were killed, many orchards were defoliated by the scale, and growers paid up to $1.00 per vedalia beetle in their eagerness to re-establish the scale predator.

Stethorus punctillum (Coleoptera: Coccinellidae)

Stethorus punctillum Weise belongs to a moderate-sized family of lady beetles (a. k. a., ladybirds or ladybugs) containing approximately 3,500 species worldwide, with almost 500 species in North America. The cosmopolitan genus *Stethorus* contains approximately 65 species; six species occur in America north of Mexico. The adults and larvae of all species in the genus *Stethorus* are specialized predators of mites. *S. punctillum* is of European origin. Therefore, it is an exotic species in North America. Its historical distribution extends from west Europe to Japan. *S. punctillum* first arrived from Turkey in 1955 for biological control of citrus red mite and avocado brown mite in orchards in southern California, USA, but did not become established. Unintentional introductions of *S. punctillum* have resulted in its establishment in southern Canada, the west coast of the United States, and northeastern United States. The distribution of *S. punctillum* probably overlaps with the distribution of a related native species, *Stethorus* punctum picipes Casey along the west coast.

Appearance

Stethorus adults are very small, hairy, oval-shaped black beetles with yellow mouthparts and antennae, with variable amounts of yellow to brownish-yellow coloration on legs. All six North American species, including *S. punctillum*, are no more than a few millimeters in total body length as adults. Total body mass of an adult ranges from 400 to 500 micrograms. *S. punctillum* pupae are also hairy and black (see image below) and larvae are grayish in color, rather slow moving with conspicuous legs (see image below). Larvae crawl from plant to plant in search of mites. Eggs are oval shaped and yellowish in color (see image below). In contrast to many other lady beetles that deposit their eggs in clusters, *S. punctillum* deposit their eggs singly, on the underside of leaves of plants infested with spider mites.

Pests Attacked (Host Range)

Under field conditions, *S. punctillum* occurs in orchards, tree plantations, gardens, and crop fields where it attacks two-spotted spider mite (*Tetranychus urticae* Koch) and other mites in the family Tetranychidae. It attacks spider mites in greenhouses, interiorscapes, nurseries and orchards. Examples of pests, other than *T. urticae*, attacked include European red mite, *Panonychus ulmi* (Koch), spruce spider mite, *Oligonychus ununguis* (Jacobi), and southern red mite, *Oligonychus ilicis* (McGregor).

Life Cycle

Adult females deposit their eggs on the underside of leaves of plants infested with mites. Newly emerged adult females require a period of approximately one week before they begin laying eggs, and only if an adequate density of spider mites is available. Adult females emerge with no or just a few mature eggs in their ovaries, thus continuous feeding on preferred prey is necessary for egg maturation. For example, females must consume at least 20 spider mites per day to commence and sustain egg production. Some reports suggest that adults prefer eating spider mite eggs rather than larval, nymphal or adult stages. Females lay from 3-13 eggs per day. The total egg production can exceed 1,000 eggs over a lifespan of up to two years, under field conditions (and overwintering). Laboratory or greenhouse cultures of *S. punctillum* produce no more than 150 eggs (on average) when fed spider mites, of all life stages, during a lifespan of less than 60 days without any winter storage. The egg stage lasts three to five days, depending on rearing temperature. The egg stage and the first instar larval stage experience the highest mortality, for reasons that are not clear. Larvae feed and develop through four instars within eight to nine days. Note that a single larva can consume 240 spider mites during its development to the pupal stage. Mature larvae move toward the lower portions of the plant to pupate, most often on the underside of leaves. Pupae require 6 - 8 days to emerge as adults. The entire life cycle takes two weeks at 26ºC. The optimum temperature for rearing *S. punctillum* ranges from 23 to 25ºC.

Relative Effectiveness in IPM Systems

In integrated pest management (IPM) systems for suppression of spider mites, *S. punctillum* is compatible with predatory mites. This predator is effective in IPM systems to control spider mites on peppers and cucumbers, but not tomatoes, in greenhouses at temperatures ranging from 20 to 30ºC. Although not determined experimentally, ineffectiveness on tomato may be due to trichomes on tomato leaves, which could ensnare *S. punctillum* larvae. Additionally, cultural practices for greenhouse tomatoes involve aggressive pruning, which could dislodge any *S. punctillum* larvae and pupae on plants during this process.

Stethorus punctillum is effective at a relative humidity from 30 to 90%. Combinations of predators, the specialist lady beetle, *S. punctillum*, and generalist predatory mite, *Amblyseius californicus* McGregor, are more successful at reducing spider mites (*T. urticae*) than either species used alone. Genetically engineered crops, such as maize, containing insecticidal proteins, derived from the microbe *Bacillus thuringiensis* Berliner, have little or no negative effects on the life history and effectiveness of *S. punctillum*. However, all host plants are not suitable substrates for *S. punctillum* larvae. Lima bean *Phaseolus lunatus* L. leaf trichomes impale larvae, reducing survival rate. Scarlet runner bean, *Phaseolus coccineus* L., leaf trichomes tear through the integument of *S. punctillum* larvae. Detailed knowledge of the interactions of host plants (crops), spider mites, and *S. punctillum* is required to increase the use of this predator in IPM systems. Another important concern is the compatibility of *S. punctillum* releases with pesticides used to control a variety of pests in IPM systems.

Pesticide Compatibility

Stethorus punctillum adults and larvae are compatible with fenbutatin (Vendex®), which is an organotin miticide, for added control of spider mites in "hot-spots" at normal field rates. Overuse of this product will reduce *S. punctillum* density on plants through a reduction in their food supply (spider mites). Pyrethroid and organophosphate insecticides can alter the seasonal activity of *S. punctillum*. Fenoxycarb (Insegar® DG), which is a carbamate insect growth regulator, and methoxyfenozide (Intrepid® 2F), which is a diacylhydrazine insecticide that induces premature and incomplete molting in lepidopteran larvae, are both known to have adverse effects on the development, emergence, and reproduction of *S. punctillum*. Other products such as soaps, supreme oils, and spreader-stickers can be harmful to *S. punctillum* if they come into direct contact with the sprays. These products have very little residual activity against *S. punctillum*. There is evidence that *S. punctillum* can develop resistance to organophosphate insecticides (e.g., azinphosmethyl) under field conditions.

Stethorus punctum (Coleoptera: Coccinellidae)

This species, first identified in 1852, is native to North America and is found throughout the mid-Atlantic region and also in the Midwest, although it is not as predominant there as in the mid-Atlantic region.

Appearance

S. punctum is a tiny (1.5 mm), oval, convex, shiny black lady beetle, covered with sparse, fine yellowish to white hairs. The white, oval eggs are less than 0.5 mm long, and turn blackish just before the larvae emerge. They are laid singly, usually on the underside of leaves near the primary vein, on their sides and adhere tightly to the leaf. The newly hatched larva is gray to blackish and has many long-branched hairs and black patches. As it matures, it becomes reddish, at first on the edges of the body, and just prior to pupation the entire larva turns reddish. The larva has 13 segments, plus the head. The pupae are black and flattened, somewhat pointed on the posterior end, with the entire body covered with yellow hairs.

Emerging adults are reddish-orange for a few hours after emerging from the pupal case before turning black. Adults can be found in the leaf litter immediately surrounding the trunks of fruit trees, in large numbers along the herbicide strip (the area under the tree canopy that is often cleared by use of herbicides) and occasionally in the drive row.

Habitat

Fruit orchards, strawberry fields, crops attacked by spider mites.

Pests attacked

Stethorus punctum is strictly a predator of plant-feeding mites, particularly the spider mites such as the European red mite and the twospotted spider mite, and especially the eggs.

Life cycle

S. punctum overwinters in the adult stage beneath leaves and other organic matter under fruit trees and in other protected habitats near the orchard, such as fence rows or adjacent wooded areas.

Adults begin to emerge at the tight cluster stage of apple development with peak emergence from the pink to bloom stage. Emergence is usually complete by the petal fall stage. Adults remain active in the orchard until September to late October.

Egg laying occurs in most areas from May to mid-August. Eggs are laid close to the primary veins of the leaf and adhere tightly, with 95 percent on the undersurface of the leaf and 5 percent on the upper surface. The larva emerges after approximately 5 days.

The larva passes through four larval stages in about 12 days, feeding on all stages of mites. The mature larva fastens itself to the leaf and remains in a motionless state for 24 to 48 hours before pupation which lasts about five days.

The peak periods of larval activity are dependent upon mite populations. There are usually three overlapping generations per year. The average period from egg deposition to the appearance of the adult is 23 days, and the adults feed for an average of 25 days before beginning to lay eggs. Females lay 1 to 10 eggs per leaf depending on mite density.

Relative Effectiveness

S. punctum is one of the most important and frequent predators of spider mites in fruit orchards. Beetles consume all stages of mites; adults can consume 75 to 100 mites per day and large larvae can devour up to 75 mites per day, so they quickly lessen an outbreak of spider mites.

Adults are very active when in fruit trees and if disturbed they will often fall to the ground. They are good fliers, and therefore tend to concentrate in areas of the orchard where mites are plentiful and disappear when the mite population becomes low. There must be 2-5 motile mites per leaf to keep *S. punctum* in an orchard, and "pockets" of 8-10 mites per leaf are required for reproduction.

Conservation

Because adults overwinter in the leaf litter immediately surrounding the trunks of fruit trees, it is advisable not to disturb the area under the tree from November 1 to mid- to late April. Recent studies indicate the distribution of *S. punctum* to be closely associated with the leaf litter distribution in the orchard.

Pesticide Susceptibility

In one study, fenoxycarb was ovicidal in the laboratory and disrupted the larval-pupal molt in the field. Abamectin was toxic to *S. punctum* larvae and adults in the laboratory and methomyl was toxic to adults. Teflubenozuron was toxic to the pupal stage in the field; along with fenoxycarb, this IGR caused late-season increases of phytophagous mite populations in field trials. Tebufenozide was not toxic to all *S. punctum* stages in the laboratory and field. *S. punctum* was tolerant to all organophosphate insecticides tested.

Lebia grandis (Coleoptera: Carabidae)

Lebia grandis belongs to a large family of beetles containing approximately 40,000 species. The cosmopolitan genus *Lebia* contains approximately 450 species. Forty-eight species occur in North America north of Mexico. The life history is known for only two of the North American species. The adults are predators, as with many carabid beetles, but larvae are ectoparasitoids of pupae or prepupae (mature larvae) of chrysomelid beetles.

Lebia grandis is the largest species in its genus found in North America. It is about one centimetre long with a rusty orange head, thorax and legs, a black abdomen and an iridescent blue or purple lustre to the dark coloured elytra. These are wide but rather shorter than the abdomen which protrudes posteriorly.

Research

The historical records for *Lebia grandis* show that it was present in areas where *L. decemlineata* was not found. This would seem to imply that at one time it parasitized a different species.

The original host plant of the Colorado potato beetle, *L. decemlineata* was the buffalo bur, *Solanum rostratum*, in Mexico. No association has been found between *Lebia grandis* and *L. decemlineata* on this host plant. In the 1850s, *L. decemlineata* began to feed on the potato plant and spread rapidly eastwards on potato crops. *Lebia grandis* may have had a relationship prior to this with the false potato beetle, *Leptinotarsa juncta* which feeds on horsenettle, *Solanum carolinense*. Another closely related chrysomid beetle, *Leptinotarsa haldemani* is recorded from Texas, Oklahoma, Arizona and Mexico, where it lives on wild species of *Physalis* and *Solanum douglasii*, but its range does not overlap with *Lebia grandis* except possibly in Texas.

In an attempt to find the original species associated with *Lebia grandis*, research was undertaken examining its prey preferences and the suitability of the three chrysomelid species, *L. decemlineata*, *L. juncta* and *L. haldemani* as hosts for its larvae.

Adult *Lebia grandis* beetles were offered suitable sized larvae of these three chrysomelid species and the number of larvae of each species that were eaten was noted. A similar experiment was performed using the eggs of the three chrysomelid species. The *Lebia grandis* adults showed no significant difference in the number of larvae of each prey species they chose to eat, but they did differentiate between their eggs. *L. juncta* egg consumption averaged nearly twice as high as *L. decemlineata* egg consumption, with *L. haldemani* being in an intermediate position.

In another experiment, a first instar larva of *Lebia grandis* was placed in close proximity to a 4th instar chrysomelid larva that was about to pupate. The success rate of parasitism in each of the three prey species was noted. Successful development of *Lebia grandis* from first instar larva to adult occurred with each of the three species of *Leptinotarsa*. However the success rate recorded for *L. juncta* was about 43% as compared to 12% for the other two species.

These studies would seem to support the view that *Lebia grandis* was originally exclusively a parasitoid of *L. juncta*, but that the opportunity afforded by its encounter with *L. decemlineata* less than 150 years ago enabled it to exploit this new and abundant food source.

Appearance

Lebia beetles are colorful as adults and range in size from 2.5 to 14 mm in length, depending on the species. *Lebia grandis* is the largest species in the genus in North America. Its body length ranges from 8.5 to 10.5 mm. Its head is usually pale (with a reddish tinge) as are its mouthparts, antennae, and thorax. Its abdomen is mostly black with a metallic blue, purple, or sometimes greenish luster to the elytra (wing covers). Its legs are entirely pale with a reddish tinge.

Lebia grandis first-instar larvae are pale to tan in coloration, with well developed appendages, mouthparts and antennae, as is typical for carabid larvae. The body length ranges from 3 to 4 mm and the width is approximately 0.5 mm. Second instar larvae undergo a gradual degeneration of appendages, develop a distended body with much reduced sclerotization (a simple form of hypermetamorphosis), eventually bearing little resemblance to the first instars.

Habitat

Lebia grandis is distributed in the eastern to mid-western United States and into adjacent Canada. It has been found inhabiting agricultural lands and their surroundings. It has been found on cultivated potato (*Solanum tuberosum*) and on horsenettle (*Solanum carolinense*) on arable land and neighboring open fields. Adults also occur on goldenrod (*Solidago* spp.).

Pests Attacked

Lebia grandis is an indigenous natural enemy of the Colorado potato beetle, *Leptinotarsa decemlineata*, and the false potato beetle, *Leptinotarsa juncta*, which is an occasional pest of eggplant. In fields of cultivated potato and eggplant, adults are specialist predators of all immature stages of Colorado potato beetle. In no-choice feeding trials in the laboratory, *L. grandis* adults consumed the larvae of the asparagus beetle (*Crioceris asparagi*), and also the three-lined potato beetle (*Lema trilinea*) but this has never been observed in the field. *L. grandis* larvae are specialist ectoparasitoids of Colorado potato beetle and false potato beetle prepupae (mature larvae) and pupae in the soil.

Lebia grandis has not been found in association with Colorado potato beetle on this pest's ancestral host plant (*Solanum rostratum*) in central Mexico. Because it was described from North Carolina in 1830, it is clear that *L. grandis* was historically a specialist enemy of the closely related false potato beetle on horsenettle in the southeastern United States, before the Colorado potato beetle adapted to potato and spread into the eastern US. Subsequently, *L. grandis* adopted the Colorado potato beetle as a new and more abundant host, and is now found as far north as Michigan and southern Maine, north of the range of false potato beetle.

Life Cycle

Adults are primarily nocturnal, but may also be seen during the day under ideal conditions of high humidity and high temperature in late spring and summer in Maryland, USA. Unusual for a temperate-zone carabid beetle, they are adapted to foraging on foliage, as opposed to running along the soil surface.

Adults emerge in late May to early June in Maryland, several weeks after the spring emergence of Colorado potato beetles. This ensures that prey (eggs and early instar larvae of Colorado potato beetle) are available for *L. grandis* adults, especially females, to feed on. It also provides adequate time for females to mate, then oviposit into soil near the base of the potato plants. Eggs are deposited singly into sandy soil. An adhesive substance (possibly a secretion from the female's accessory glands) covers each egg as it is laid, causing each to adhere to sand granules and become difficult to detect. A single female *L. grandis* can lay as many as 1300 eggs in its lifetime.

As first instar *L. grandis* emerge from the egg stage (within several days), they are very sensitive to dryness, but fairly-well resistant to wet conditions. They readily search in the soil for Colorado potato beetle larvae about to pupate. First instars may follow an odor trail left behind by the Colorado potato beetle mature larvae, which burrow into the soil to pupate. How quickly the larvae must find it host, in order to insure successful parasitism, may depend on soil conditions which allow the host to seal its pupal cells.

Once locating the host, the first instar larva attaches to the integument (skin) of the host with its mandibles and begins feeding, all of which takes place during several days of the first larval stadium, killing the host. After molting, the second instar *L. grandis* larva does not resume feeding. Metamorphosis to the pupal stage occurs soon thereafter, without any period of diapause. At 25°C, the adult emerges from the soil in 28 days from when it began feeding on its host; below this temperature emergence decreases, and at 17°C, only 2.5% of adults emerged after a mean of 73 days. No adults developed above a constant temperature of 33°C, for which the mean developmental time was 19 days. Two generations of *L. grandis* are probably produced each year in many populations in the southeastern United States. Adults overwinter beneath the soil surface in or near potato fields.

Effectiveness

L. grandis have been considered by some to be the most promising indigenous enemy of Colorado potato beetle in North America. No large scale field studies have been conducted to date. However, cage studies support its ability to reduce pest numbers. Under field conditions, *L. grandis* could be an effective predator/parasitoid of Colorado potato beetle on potato or eggplant, when used in combination with other control strategies.

Foliar applications of *Bacillus thuringiensis* tenebrionis are compatible with the action of *L. grandis*. However, natural densities of *L. grandis* will not be great enough to effect control of this pest. Thus, augmenting populations by releasing mass reared adults is one potential method for maximizing the effectiveness of this enemy of Colorado potato beetle. However, rearing experiments have resulted in only limited success in mass-producing *L. grandis*.

Conservation

In fields of normal (non-transgenic) potato, the judicious use of pesticides will help conserve carabid populations. Timing of pesticide applications prior to spring emergence of adults would also help reduce unintentional killing of these natural enemies.

In transgenic potatoes (containing the delta endotoxin derived from the bacterium, *Bacillus thuringiensis* ssp. *tenebrionis*) *L. grandis* adults will not persist due to the low availability of prey to feed on and larval hosts to parasitize.

Presence of *L. juncta* as alternative prey and hosts, either on horsenettle or eggplant (both are hosts of both species), will tend to conserve *L. grandis* populations in the absence or scarcity of Colorado potato beetle.

Pesticide Susceptibility

As far as known, *Lebia grandis* adults and larvae can be killed by organophosphate, carbamate, or pyrethroid insecticides when contacting residues on the ground or foliage of potato plants. However, spray formulations of Bacillus thuringiensis tenebrionis (Btt or Bt San Diego) are not harmful.

Aleochara bilineata (Coleoptera: Staphylinidae)

The rove beetles are the largest North American beetle family. Most are predaceous, and they are common in organic matter. *Aleochara bilineata* as an adult is predaceous, but its larvae are parasitic.

Appearance

A. bilineata adults are glossy black with short coarse hairs and are about 5-6 mm long. They have very small, reddish-brown, inconspicuous forewings beneath which the membranous hindwings are elaborately folded. The forewings look like pads on the upper abdomen. The long abdomen is held aloft like a scorpion if the beetles are disturbed. Well-developed "jaws" cross in front of the head of both adults and larvae. Adults are active, strong fliers.

Eggs are oval, about 0.5 mm long and 0.4 mm wide, and are covered with a gelatin-like material which is pale green at first and turns darker after a few days. First instar larvae are pale brown, about 1.5 mm long, slender, segmented, and tapered toward the anterior. They have large heads. The parasitic second and third instar larvae are white, have rudimentary legs, and are found within the host puparium.

Habitat (Crops)

Cole crops, onions, turnips, radish, corn, and others.

Pests Attacked

Root maggot eggs, larvae, and pupae, especially the cabbage and onion maggot.

Life Cycle

A. bilineata overwinters as a first instar larva within a host puparium. Two days after mating, females begin laying tiny, whitish elliptical eggs—about 15 per day with an average of 700 per season. Eggs are deposited in the soil among the roots of root maggot-infested plants. Larvae hatch in about 5 to 10 days and actively search for root maggot puparia in the surrounding soil. The larva punctures the host puparium, enters, and begins feeding. The entire host pupa is consumed. Although two or more larvae may occasionally enter the same puparium, only one will survive to maturity.

A. bilineata pupates within the host puparium and emerges as an adult after 30 to 40 days. The overwintering larvae emerge in the second half of July, and the second generation emerges at the end of August or in early September. Adults live 40 to 60 days, and the life cycle, from egg to adult is about 6 weeks. There may be two generations per year.

Relative Effectiveness

A. bilineata adults may consume up to five root maggot larvae per day—a pair may destroy 1200 eggs and 130 larvae, and during their lifetime their offspring may parasitize several hundred pupae. Adults are cannibalistic, eating their own eggs and attacking other adults when food supplies are low.

Adult rove beetles may not emerge in spring until several weeks after the overwintering root maggot adults emerge. This is too late to suppress early season crop damage, although late season parasitism of cabbage maggot pupae may sometimes be as high as 90% to 95%. In one field study, parasitism ranged between 30% and 70%.

A. bilineata has been mass reared in the former Soviet Union, Europe, and Canada for control of root maggots. It may therefore be feasible to mass release adults at the beginning of the season to increase predation and parasitism of the first generation of root maggots.

Pesticide Susceptibility

Laboratory tests showed *A. bilineata* to be highly susceptible to pyrethroid insecticides. A strain of *A. bilineata* selected for tolerance to cyclodiene-type insecticides proved successful as a natural enemy of cabbage maggots in Canadian trials in the 1960's. In a 1980 field study, the insecticide chlorfenvinphos used in conjunction with *A. bilineata* successfully controlled cabbage maggots, although parasitism was reduced.

Conservation

Rove beetles are found under debris and rocks, near water, in compost and piles of decaying material, or in the crop canopy. The adults have been found in sweet corn tassels and silks late in the season. Hedgerows and shelter belts may provide protection for adult beetles.

BUGS

Campylomma verbasci (Meyer) (Hemiptera: Miridae)

MULLEIN PLANT BUG

Campylomma is a generalist predator of apple and pear orchard pests including mites, aphids, and pear psylla. Unfortunately it is also recognized as a pest of apple fruit and in rare instances may cause damage to pear. Adults and nymphs are predacious, but may feed on fruit (causing cosmetic damage to skin of fruit) if available prey are reduced to very low numbers. *Campylomma* occurs in most deciduous fruit growing regions of the northern U.S. and southern Canada.

Appearance

The adult is green-brown, elongated oval in shape, and about 1/10 inch (2.5mm) long. It has a dark spot on the first antennal segment and black spines on the legs. The nymphs are ovate and translucent when first hatched, but gradually turn pale green. The egg is about 1/28 inch (0.87mm) long and sac-shaped. It is inserted into the bark, stems, and/or leaves of host plants with only the operculum (cap or cover) exposed.

Bugs

Habitat

Campylomma is found in both pome fruit trees and herbaceous plants, particularly mullein.

Pests Attacked

Common prey include aphids, mites, thrips, and pear psylla.

Life Cycle

Campylomma overwinters as an egg in apple and pear bark and perhaps other woody deciduous hosts. Eggs hatch in the spring before bloom of apple and pear. Nymphs develop through five instars in about 21 days at 72 °F (21°C). The period of nymphal development of the first generation is the time fruit is most likely to be damage by *Campylomma* feeding. Adults first appear in mid to late May in the Pacific Northwest and a portion of the population moves into surrounding herbaceous hosts, particularly mullein (hence the common name), where they feed on thrips and other available prey. They migrate back into the orchards in late summer, where they mate and lay overwintering eggs. There are from two to four generations per year in the Pacific Northwest.

Relative Effectiveness

Campylomma can have a major impact on pear psylla populations in pear and aphid populations in apple. This mirid predator appears to be tolerant to many insecticides and is one of the few predators routinely found in heavily sprayed orchards.

Pesticide Susceptibility

Certain pesticides are known to be highly toxic to *Campylomma*, chlorpyrifos and formetanate hydrochloride are used when control is necessary, while others may have suppressive or little effect.

Conservation

Campylomma must be monitored closely early in the season in order to distinguish between beneficial and potentially damaging populations. Although apparently tolerant of many pesticides, use of pesticides that have a narrow spectrum of activity will help conserve this predator.

Deraeocoris nebulosus (Uhler) (Heteroptera: Miridae)

Deraeocoris nebulosus (Uhler) is a generalist predator of plant-feeding insects and mites. It is associated with many common pests on more than 50 species of ornamental trees and shrubs. *D. nebulosus* is found in southern

Canada and is widespread in the United States; it is common in the eastern states. Reported by the researcher to be predaceous on a cankerworm, it may have been the first mirid documented as a predator in North America.

Appearance

Adults have ovate, shiny, dark, olive bodies with pale markings, and are 3.5-4.0 mm long and 1.75-2.0 mm wide. The apical half of each forewing is clear with each having a small fuscous dot which helps distinguish *D. nebulosus* from other species in the genus. Eggs (0.91 mm long, 0.28 mm wide) are laid singly or in groups of two or more in leaf midveins and petioles, with only the operculum, including a respiratory horn, visible. Nymphs are pale grey, with the early instars having a red tinge and the late instars having red streaks on the legs and a red line between two of the segments on the abdominal dorsum.

Habitat

Deraeocoris nebulosus, collected on more than 75 plant species, is found in apple orchards, peach orchards, pecan orchards, cotton fields and in landscape plantings.

Pests Attacked (Host Range)

This predator feeds on whiteflies, aphids, psyllids, scales, mites, and lace bugs. In captivity it tends to cannibalize unless provided with hiding places.

Life Cycle

Adults overwinter in protected places such as under bark. With onset of warmer weather and appearance of new leaves, eggs are laid in the petioles and midveins. This bug has five nymphal stages. Under laboratory conditions, the nymphal period lasts 19.8 days at 21°C and 13.3 days at 27°C. Females have a mean fecundity of about 240 eggs. Three generations per year are reported from Pennsylvania, with more suspected farther south. Adults track a succession of plant species, according to prey availability.

Relative Effectiveness

Researchers reported that *D. nebulosus* consumed an average of 107.6 lace bug nymphs during development and 6.9 nymphs per day as an adult. An adult can consume 4-7 cotton aphids per day and 16-19 tobacco budworm eggs per day. Little work has been done with *D. nebulosus* in the field.

Conservation

Use of target-specific pesticides only after a pest reaches an economic threshold will help conserve *D. nebulosus*.

The conservation of natural enemies is probably the most important and readily available biological control practice available to growers. Natural enemies occur in all production systems, from the backyard garden to the commercial field. They are adapted to the local environment and to the target pest, and their conservation is generally simple and cost-effective. With relatively little effort the activity of these natural enemies can be observed. Lacewings, lady beetles, hover fly larvae, and parasitized aphid mummies are almost always present in aphid colonies. Fungus-infected adult flies are often common following periods of high humidity. These natural controls are important and need to be conserved and considered when making pest management decisions. In many instances the importance of natural enemies has not been adequately studied or does not become apparent until insecticide use is stopped or reduced. Often the best we can do is to recognize that these factors are present and minimize negative impacts on them. If an insecticide is needed, every effort should be made to use a selective material in a selective manner.

Pesticide Susceptibility

Little work has been done on the susceptibility of *D. nebulosus* to pesticides. Synthetic pyrethroids show a low toxicity, only up to 10% and 30% susceptible at the low and high rates tested, respectively (Croft and Whalon 1982).

Deraeocoris brevis, a species common in the northwestern United States, is susceptible to many pesticides including azinphosmethly, fenvalerate, diflubenzuron, and organophosphates (Westigard 1973, van de Baan and Croft 1990, Booth and Riedl 1996). *D. nebulosus* probably is also susceptible to these pesticides.

Geocoris spp. (Heteroptera: Lygaeidae)

BIGEYED BUG

Bigeyed bugs, *Geocoris* spp., are among the most abundant and important predaceous insects in many cropping systems in the U.S. They are also found in lawns and ornamental plantings. There are approximately 19 species that inhabit North America. Of these, *G. punctipes* and *G. pallens* are the most common. *Geocoris* spp. are known to feed on plants; however they rarely cause economic damage.

The most common species found in North America are:

- *G. punctipes*
- *G. pallens*
- *G. bullatus*
- *G. uliginosus*

Appearance

Adults and nymphs have oval bodies and broad heads. Their most distinguishing characteristic is their large, bulging eyes. They have relatively short antennae that are slightly enlarged at the tip. Bigeyed bugs walk with a distinctive "waggle" and omit a fowl odor when handled. Adults are about 3/16th inch long and silver/gray in appearance (*G. punctipes*). The immatures look similar to small adults, but lighter and without fully developed wings. "Hot dog"-shaped eggs are laid singly on leaves (typically on undersides) and stems of many crops. They are white to tan with a distinctive red spot. Both the adults and immatures feed by sucking juices from their prey through a needle-like beak. The beak is folded under the insect's body when it is not feeding.

Beware: chinch bugs, a pest of turfgrass, resemble bigeyed bugs and managers sometimes mistakenly spray when the biological control agent is found.

Habitat

Geocoris spp. are commonly found throughout the southern regions of the U.S. in most cropping systems. *G. punctipes* and *G. pallens* are the most common species along the cotton belt. *G. uliginosis* is frequently found in lawns. *G. punctipes* is also found in ornamental plantings and vegetables. Look among leaf litter at the base of plants to detect bigeyed bugs. *G. punctipes* is frequently found in strawberries and vegetables grown in plasticulture in GA and NC.

Pests Attacked

Bigeyed bugs feed on a wide variety of prey smaller than themselves. They are among the most important natural enemies in cotton. They feed on eggs and small larvae of most lepidopteran pests (bollworm, pink bollworm, tobacco budworm), on the eggs and nymphs of plant bugs (e.g., lygus), and on all life stages of whiteflies, mites and aphids. There are 67 species included in a list of prey organisms fed on by bigeyed bugs. Researchers observed that *Geocoris* use salivary secretions to adhere whitefly wings to a stable surface to allow feeding.

Life Cycle

Bigeyed bugs overwinter as adults. As with any insects, life cycle is dependent on temperature. Eggs that are laid singly on leaves or stems hatch in approximately one week. Bigeyed bugs have five nymphal instars, each of which lasts from 4 to 6 days. Both nymphs and adults are predaceous. Adults live approximately one month and a female can lay up to 300 eggs during her adult life span. Several generations may occur during a crop growing season.

Relative Effectiveness

The potential for *Geocoris* spp. as biological controls is well founded. Adults and immatures can consume dozens of prey per day. However, due to the small size and cryptic nature of this beneficial, quantification of predation in the field is difficult. Cover crops such as crimson clover and vetch may enhance populations of bigeyed bug and improve predation against pests in some crops. Other studies have demonstrated the potential for effectiveness against various pests in greenhouse crops.

Conservation

Geocoris spp. are ideal candidates for conservation. Since bigeyed bugs are also seed feeders, refuge crops can be planted that harbor large populations of *Geocoris*. For example, researchers investigated the possibility of increasing bigeyed bug populations in crop lands by planting sunflowers. Their results were very encouraging. It is suggested that the ability to obtain some nutrition from plants further increases bigeyed bug potential as a useful biological control agent because it provides feeding flexibility during times of low pest populations, allowing populations to remain stable and ready to address further pest outbreaks. Exploiting the potential of *Geocoris* spp. as a native biological control agent deserves further study.

Polycropping, use of pest-specific insecticides (e.g., microbials, insect growth regulators, Bt cotton, etc.), and use of economic thresholds to minimize pesticide applications are all practical recommendations for preserving *Geocoris* (and other natural enemies) populations.

Pesticide Susceptibility

As with most beneficial insects, *Geocoris* spp. are very susceptible to broad spectrum pesticides. Conservation of bigeyed bugs and the use of broad spectrum pesticides are not compatible. Because bigeyed bugs are known to feed on plants, systemic pesticides may have negative effects on *Geocoris* populations. Researchers showed that leaf residues of foliar applications of some insecticides were lethal to bigeyed bugs. Among a variety of cotton insecticides, Elzen, Elzen and King (1998) found that cyfluthrin, profenofos, endosulfan, oxamyl and spinosad were most favorable to conservation of bigeyed bugs, while malathion, chlorfenapyr and fipronil caused the most *Geocoris* mortality. Similar results were found in another study of insecticide residues in soybean.

Orius tristicolor and *O. insidiosus* (Hemiptera: Anthocoridae)

MINUTE PIRATE BUG AND INSIDIOUS FLOWER BUG

Orius spp. are "true" bugs and occasionally may bite humans, but the bite is only temporarily irritating.

Appearance

Adults are very small (3 mm long), somewhat oval-shaped, and black with white wing patches. Wings extend beyond the the tip of the body. Nymphs are small, wingless insects, yellow-orange to brown in color, teardrop-shaped and fast moving. Both adults and nymphs feed by sucking juices from their prey through a sharp, needle-like beak (the rostrum), which is characteristic of all true bugs.

Habitat

Orius is common on many agricultural crops including cotton, peanuts, alfalfa, corn, pea, and strawberry, on pasture land, in orchards, and is successfully used as a biological control agent in greenhouses. It is often found in corn silks and is most common where there are spring and summer flowering shrubs and weeds since it feeds on pollen and plant juices when prey are not available. In the Midwest, *O. insidiosus* is more common, while *O. tristicolor* is more common in the western states. In greenhouses, *Orius* spp. are used as generalist predators, especially on cucumber and bellpepper crops.

Pests attacked

Both immature stages (nymphs) and adults feed on a variety of small prey including thrips, spider mites, insect eggs, aphids, and small caterpillars. *Orius* holds its prey with its front legs and inserts its beak into the host body, generally several times, until the soft body is empty and only the exoskeleton remains. It has been reported to be an important predator of the eggs and new larvae of the bollworm and of spotted tobacco aphid, but it is believed that thrips and mites are the more basic part of an *Orius* diet. It can also be an important predator of corn earworm eggs which are laid on the silks. Other reported prey include eggs and small European corn borers, corn leaf aphids, potato aphids, and potato leafhopper nymphs.

Life cycle

Females lay tiny eggs 2-3 days after mating within plant tissues where they are not easily seen. These hatch into nymphs which develop through five nymphal stages. Egg incubation is generally 3-5 days, and development from egg to adult takes a minimum of 20 days under optimum conditions. Females lay an average of 129 eggs during their life spans, and adults live about 35 days. Several generations may occur during a growing season.

Relative Effectiveness

Both immature and adult bugs can consume 30 or more spider mites per day, although *Orius* has been observed to leave prey before having completely consumed it to attack another mite. Nevertheless, the first mite was incapacitated and so more mites may be destroyed than those needed to fulfill nutritional requirements.

Conservation

Diversified cropping systems, use of microbial insecticides, e.g., products containing *Bacillus thuringiensis,* and use of economic thresholds to minimize insecticide applications, are all practical recommendations to maximize the natural biological control from *Orius*. Plantings of spring and summer flowering plants will help them survive periods of scarce prey.

The conservation of natural enemies is probably the most important and readily available biological control practice available to growers. Natural enemies occur in all production systems, from the backyard garden to the commercial field. They are adapted to the local environment and to the target pest, and their conservation is generally simple and cost-effective. With relatively little effort the activity of these natural enemies can be observed. Lacewings, lady beetles, hover fly larvae, and parasitized aphid mummies are almost always present in aphid colonies. Fungus-infected adult flies are often common following periods of high humidity. These natural controls are important and need to be conserved and considered when making pest management decisions. In many instances the importance of natural enemies has not been adequately studied or does not become apparent until insecticide use is stopped or reduced. Often the best we can do is to recognize that these factors are present and minimize negative impacts on them. If an insecticide is needed, every effort should be made to use a selective material in a selective manner.

Pesticide Susceptibility

Foliar applications of insecticides to crops can greatly reduce *Orius* numbers. Even soil-applied systemic insecticides may reduce their numbers because of their habit of sucking plant juices.

Podisus maculiventris (Hemiptera: Pentatomidae)

SPINED SOLDIER BUG

One of the stink bugs, named for the strong odor emitted when disturbed, this useful insect has a wide host range, including several important crop pests. It is the most common of several species of *Podisus* and is found throughout the continental United States.

Appearance

Adult spined soldier bugs are pale brown to tan and can be 8.5-13 mm long. They are shield-shaped with prominent spurs on the "shoulders," immediately behind the head. Unlike pest species that may look similar, spined soldier bug adults have a distinctive dark line on the membranous tip of each forewing that may form one dark streak when the wing tips overlap.

Young nymphs are red and black; older nymphs are marked with red, black, yellow-orange, and cream bands and patches. They are wingless and round rather than shield-shaped. Both adults and nymphs have long, pointed beaks with which they stab their prey and which they keep folded under their bodies when not feeding.

Habitat (Crops)

Most, including potatoes, tomatoes, sweet corn, cole crops, beans, eggplant, cucurbits, asparagus, apples, and onions.

Pests Attacked

Over 100 species in many families have been reported as prey. Prime targets are immature insects. Reported prey include the larvae of Mexican bean beetle, European corn borer, diamondback moth, corn earworm, beet armyworm, fall armyworm, cabbage looper, imported cabbageworm, Colorado potato beetle, velvetbean caterpillar, and flea beetles.

Life Cycle

Each female lays several hundred gray, cream, or gold barrel-shaped eggs in tight clusters of 20 to 30 on leaves and twigs. The nymphs initially cluster around the hatched eggs, then disperse to feed. There may be two to three generations per year. In the laboratory, adults have lived 2-3 months.

Relative Effectiveness

Individual spined soldier bugs have been recorded as consuming more than 100 late instar fall armyworm larvae over the equivalent of a season.

In Washington potato fields, spined soldier bugs have been released in large numbers, with twospotted stink bugs, to reduce infestations of Colorado potato beetle by up to 50%. The species is sold for the control of Mexican bean beetle, but its effectiveness against this pest has not been proven yet in large scale trials. A pheromone, formulated to attract spined soldier bugs, is also commercially available.

Pesticide Susceptibility

Research showed *P. maculiventris* to be more susceptible than its prey, velvetbean caterpillar, fall armyworm, and corn earworm, to organophosphorous and carbamate insecticides, but less susceptible to pyrethroids.

Studies have shown that diafenthiuron and diflubenzuron, although less toxic by residual contact, were toxic to *P. maculiventris* when ingested in water and that pyriproxyfen and imidacloprid caused significant mortality to *P. maculiventris* populations regardless of the manner of contact.

Conservation

Conservation of natural enemies is arguably the most important concept in the practice of biological control and fortunately is also one of the easiest to understand. Simply put, conservation of natural enemies means avoiding practices which are harmful to natural enemies and implementing practices which benefit them. Sounds like good common sense! The tricky part comes in understanding exactly what practices are harmful and how beneficial practices can be integrated into your production system. That requires that we understand the biology of the natural enemies and are willing to modify our practices to accommodate them.

LACEWINGS

Chrysoperla (=Chrysopa) carnea, C. rufilabris (Neuroptera: Chrysopidae) Common Green Lacewing (=C. carnea)

These green lacewings are common in much of North America. Adults feed only on nectar, pollen, and aphid honeydew, but their larvae are active predators. *C. carnea* occurs in a wide range of habitats in northeastern, midwestern and western U.S., and *C. rufilabris* may be more useful in areas where humidity tends to be high (greenhouses, irrigated crops, southeastern and midwestern U.S.).

Appearance

Adult green lacewings are pale green, about 12-20 mm long, with long antennae and bright, golden eyes. They have large, transparent, pale green wings and a delicate body. Adults are active fliers, particularly during the evening and night and have a characteristic, fluttering flight. Oval shaped eggs are laid singly at the end of long silken stalks and are pale green, turning gray in several days. The larvae, which are very active, are gray or brownish and alligator-like with well-developed legs and large pincers with which they suck the body fluids from prey. Larvae grow from <1 mm to 6-8 mm.

Habitat (Crops)

Cotton, sweet corn, potatoes, cole crops, tomatoes, peppers, eggplants, asparagus, leafy greens, apples, strawberries, and other crops infested by aphids.

Pests Attacked

Several species of aphids, spider mites (especially red mites), thrips, whiteflies, eggs of leafhoppers, moths, and leafminers, small caterpillars, beetle larvae, and the tobacco budworm are reported prey. They are considered an important predator of long-tailed mealybug in greenhouses and interior plantscapes.

Life Cycle

These two species of green lacewings overwinter as adults, usually in leaf litter at the edge of fields. During the spring and summer, females lay several hundred small (<1 mm) eggs on leaves or twigs in the vicinity of prey. Larvae emerge in 3-6 days.

The larval stage has three instars and lasts two to three weeks. Mature third instars spin round, parchment-like, silken cocoons usually in hidden places on plants. Emergence of the adults occurs in 10 to 14 days. The life cycle (under 4 weeks in summer conditions) is heavily influenced by temperature. There may be two to several generations per year.

Relative Effectiveness

These lacewing larvae are considered generalist beneficials but are best known as aphid predators. The larvae are sometimes called aphid lions, and have been reported to eat between 100 and 600 aphids each, although they may have difficulty finding prey in crops with hairy or sticky leaves.

Natural populations of *Chrysoperla* have been recorded as important aphid predators in potatoes, but mass releases of lacewings have yet to be evaluated against aphids in commercial potato production. In small scale experiments outside the United States, lacewings achieved various levels of control of aphids on pepper, potato, tomato, and eggplant, and have been used against Colorado potato beetle on potato and eggplant. On corn, peas, cabbage, and apples, some degree of aphid control was obtained but only with large numbers of lacewings. Mass releases of *C. carnea* in a Texas cotton field trial reduced bollworm infestation by 96%, although more recent studies show that *C. carnea* predation on other predators can disrupt cotton aphid control.

C. carnea is considered an important aphid predator in Russian and Egyptian cotton crops, German sugar beets, and European vineyards. The North Carolina State University Center for IPM considers it an important natural enemy of long-tailed mealybug, one of the 5 most important pests of NC interiorscapes.

Several strains of *C. carnea* occur in North America. Matching of the proper strain to specific pest management situations is desirable.

Pesticide Susceptibility

C. carnea appears to have some natural tolerance to several chemical insecticides although there may be considerable variation. Populations tolerant of pyrethroids, organophosphates, and carbaryl have been selected in the laboratory.

Conservation

Because young larvae are susceptible to dessication, they may need a source of moisture. Adult lacewings need nectar or honeydew as food before egg laying and they also feed on pollen. Therefore, plantings should include flowering plants, and a low level of aphids should be tolerated. Artificial foods and honeydew substitutes are available commercially and have been used to enhance the number and activity of adult lacewings. These products may provide sufficient nutrients to promote egg laying, but they cannot counter the dispersal behavior of newly emerged adult lacewings.

Hemerobius spp. (Neuroptera: Hemerobiidae)

BROWN LACEWINGS

Brown lacewings may be less commonly observed than the green lacewings (Neuroptera: Chrysopidae), but they are also important predators of aphids and other small soft-bodied insects. Most of the Hemerobiidae look superficially alike. The most common genus occurring in the United States is *Hemerobius*. The most common species of *Hemerobius* is *H. stigma* (=*stigmaterus*) Stephens.

Appearance

Hemerobius stigma is widely distributed across most of North America and is themost common conifer-inhabiting brown lacewing in most of its range. It is relatively large for a brown lacewing (5.0-7.5 mm). The fore wings are narrowly oval with a slightly pointed apex. Color is a light reddish brown to brownish grey. A close look at the wing venation can distinguish it from most other *Hemerobius* species. Species determinations, however, can only be confirmed by expert examination of the genitalia.

Habitat (Crops)

Typically, brown lacewings become active earlier in the season than green lacewings. They tend to prefer arboreal habitats such as coniferous and deciduous forests, plantations, or orchards. *Hemerobius stigma* is particularly associated with conifers.

Pests Attacked

All brown lacewings are voracious predators as both larvae and adults. Prey includes tree-dwelling insects such as aphids, adelgids and other small soft-bodied insects. *H. stigma* is often found in association with the balsam twig aphid (*Mindarus abietinus* Koch) in Christmas tree plantations. It is also an important predator of the balsam woolly adelgid (*Adelges piceae*) and pine bark adelgid (*Pineus strobi*). Researcher recommended it for control of Cooley[1]s spruce gall adelgid (*Adelges cooleyi*) on Douglas-fir. It is known to attack a variety of conifer-feeding aphids.

Life Cycle

H. stigma overwinters as an adult or prepupa. It becomes active as early as March in Wisconsin or April in Quebec. Active adults have even been observed in December and February in British Columbia. In early spring, the adults mate and eggs are deposited about two weeks later. The adult females and the eggs are extremely cold-tolerant. Eggs are laid singly, on fir needles or between bud scales. Eggs hatch in about 11 days, depending on the temperature.

Most hemerobiids have three larval instars. All instars are active feeders. They are relatively slender, with a small head and jaws compared to chrysopid larvae. After the third instar, the larva forms a pupa in a silk cocoon. The pupal stage lasts about 9-14 days in the summer. Two generations per year are most common, although temperature can play a significant role.

Relative Effectiveness

Hemerobius species have great potential as biological control agents for several reasons: the adults are long-lived and have a high reproductive capacity, both larvae and adults have voracious appetites, and their prey includes many pest species.

Pesticide Susceptibility

As with most beneficial insects, brown lacewings are susceptible to broad-spectrum insecticides.

Conservation

Conservation appears to be the best way to encourage populations of brown lacewings in the field. Avoiding insecticide use when lacewings are present and encouraging forested windbreaks or trees along field edges may be helpful. Adults are easy to scout by foliage beating over a sheet —they have a 'drop dead' response when disturbed and will remain motionless for a short time. Use this scouting method to determine if lacewings are present and active in early spring.

FLIES

Syrphid Flies (Diptera: Syrphidae)

Syrphid flies are also called flower flies or hover flies as the adults are usually seen hovering over or feeding on nectar or pollen at flowers. The larvae can be voracious predators of small insects, especially aphids, and all aphid-feeding (aphidophagous) species are in the subfamily Syrphinae. Common aphidophagous syrphid flies in California vegetable crops include *Toxomerus marginatus, Allograpta obliqua, Syrphus opinator, Eupeodes americanus, Eupeodes volucris, Paragus tibialis, Platycheirus spp., Scaeva pyrastri, Sphaerophoria spp., Syrphus opinator,* among many others. Syrphids are found throughout North America and are common on many crops attacked by aphids and other small soft-bodied insects.

Appearance

Adult flies vary in length from 3 to 13 mm depending on species. Adult bodies are black or brown marked with bands or dots of white or yellow covering abdomen and/or thorax. Their resemblance to bees and wasps is mimicry to ward off predators. However, they can be distinguished by a single pair of wings, with vestigial hind wings as balancing organs, characteristic of all dipterans. Syrphid fly larvae are slug-like maggots that are wrinkled and tapered anteriorly. Their color can be pink, yellow, green or brown marked with white or black, and they vary in size from 4 to 18 mm in length.

Habitat

Adult flies visit flowers and feed on nectar and pollen. They are considered important pollinators for some plants. They can also be often found near aphid-infested leaves where they lay eggs and also feed on honeydew secreted by aphids. Maggots of most species are found on aphid-infested plants like citrus, subtropical fruit trees, grains, corn, alfalfa, cotton, grapes, lettuce and other vegetables, ornamentals, and many wild host plants of aphids. Larvae of some species live in various habitats such as the nests of social insects (ants, termites or bees), decaying plant or animal matter, and stagnant water (Eristalis spp.). A few feed on live plants.

Pests Attacked

Most syrphid fly maggots feed on aphids, thrips, leafhoppers and and other soft-bodied prey like small caterpillars. They move along plant surfaces, lift their heads to grope for prey, seize and suck them dry and then discard the exoskeleton.

Life Cycle

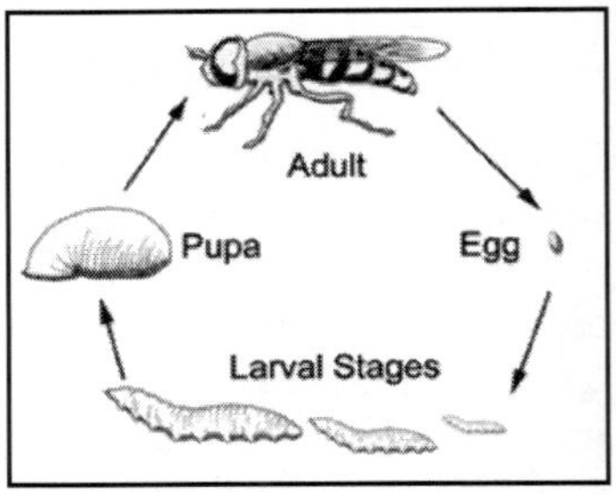

The life cycle varies among species and depends on environmental conditions and availability of food. In optimal conditions, single white eggs are laid on leaves infested with prey or other suitable food source. After hatching in about 3 days, larvae develop through several instars (usually 3) over a period of 1 to 3 weeks. The last instar turns into tan-brown teardrop-shaped puparium either on the host plant or in the soil. Unless the pupal stage remains for overwintering, adults emerge in 1 to 2 weeks. Up to seven generations can occur within a year.

Line drawing of *Toxomerus* spp. life cycle (Brett Blaauw, Department of Entomology, Michigan State University, East Lansing, Mi 48823).

Relative Effectiveness

Each larva can consume up to 400 aphids during development. When syrphid larvae are abundant, they may reduce aphid populations by 70 to 100%.

Pesticide Susceptibility

Lady beetles may be tolerant of some chemical insecticides applied at recommended field rates. But tolerance may be influenced by prior exposure of the natural enemy population to the chemical, and there is considerable variation. For example, *Coleomegilla maculata* has been shown to be susceptible to carbaryl and methamidophos at doses that might be used to control aphids in potatoes. Overwintering adults may be less susceptible to chemical insecticides than active adults and larvae.

Conservation

Applications of insecticides, especial broad-spectrum ones, should be avoided when possible. As adult hover flies feed on pollen and nectar, planting flowering plants near the field or between crop plants will attract them and increase the possibility that they lay their eggs on infested plants. Managing hedgerow plants or other vegetation may help provide alternative prey, wind shelter and even overwintering habitat.

MIDGES

Aphidoletes aphidimyza (Diptera: Cecidomyiidae)

The aphid midge, *Aphidoletes aphidimyza,* is a cecidomyiid fly whose larvae are effective predators of aphids. Initially a new species was described for every different prey attacked, and at least 24 previously described species have been determined to be *A. aphidimyza*. This midge is an important component of biological control programs for greenhouse crops and is widely sold in the U.S.

Appearance

Adult aphid midges are small (2-3 mm), delicate, mosquito-like flies with long, dangling legs and long antennae. Eggs are oval, minute (about 0.1 mm by 0.3 mm), and orange. The larvae, which look like orange maggots, are tiny, growing through three instars from minute to 2-3 mm. Depending on their food source, they are bright orange to red, and their bodies narrow toward the head. The larvae have strong "jaws" with which they grasp their prey.

Habitat (Crops)

Cole crops, potatoes, greenhouses, backyard gardens, ornamentals, orchards, berries. In the greenhouse *A. aphidimyza* has become very important for aphid control on long term vegetable crops. Although very few field introductions have been reported, studies indicate there is the potential for aphid control in many outdoor crops. *A. aphidimyza* has been found on cabbage, apples, blueberries, and ornamental bushes, and is recognized as being an important naturally occurring control agent of aphids on Russian and Egyptian cotton.

Pests Attacked

A. aphidimyza attacks over 60 species of aphids.

Life Cycle

Each female may live for one to two weeks and deposit, singly or in clusters, about 70 upright, orange eggs on leaves among aphids. The eggs hatch in two to four days. The midge larva paralyses each aphid by attacking its leg joints and then sucks it dry, leaving a blackened, collapsed aphid attached to the leaf.

In greenhousese the larvae drop to the soil less than a week after hatching and burrow to pupate; adults emerge about one to two weeks later. In the field, larvae develop in one to two weeks, and pupation may take up to three weeks. The life cycle in the field may range from three to six weeks, and there may be three to six generations per year, depending heavily on daylength.

Adult midges fly at night and are rarely seen; they feed on honeydew. Aphid midges overwinter as larvae in cocoons in the soil, pupating in spring. Adults emerge in late spring, mate the night of emergence, and the mated females begin the search for aphids. Most eggs are laid during the first few days after emergence.

Relative Effectiveness

Adult midges are very efficient at locating aphid colonies. In one study, *A. aphidimyza* located the one infested plant out of 75. One larva needs a minimum of 7 aphids in order to complete the life cycle, but it may eat as many as 80. In addition, larvae kill more aphids than they consume. Greenhouse vegetable integrated pest management programs in Canada recommend the use of these midges, often in conjunction with releases of the parasitoid wasp, *Aphidius matricariae,* for aphid control. The larvae are most common from mid- to late summer in the field. Mass released aphid midges have adequately controlled aphids in small backyard garden trials.

The onset of shorter daylength toward the end of summer can induce diapause. This can be averted in the greenhouse by nocturnal use of 100 W incandescent light bulbs spaced 22 m apart when the canopy is open, with closer spacing when the canopy is dense.

Pesticide Susceptibility

Most sprays are toxic to midges. Adults may be more susceptible than larvae.

Conservation

Aphid midges are native to much of North America and will overwinter, although winter mortality may be high. High humidity and shelter from high temperatures and strong winds will encourage midge activity. In greenhouse environments, adult midges are most effective at 20-26 degrees C with high humidity. Adults also need a source of honeydew for feeding and to improve egg laying. Larvae need slightly moist soil for successful pupation.

MITES

Apple Orchard Mite Control *Galendromus (=Typhlodromus, =Metaseiulus) occidentalis* (Acarina: Phytoseiidae)

WESTERN PREDATORY MITE

Predatory mites are important control agents of pest mites in North American orchards. The western United States native, *Galendromous occidentalis*, is a primary biocontrol agent of pest mites on many crops, but its benefit to commercial apple, plum, peach, and cherry orchard farming is especially well known.

Appearance

Adults are pear-shaped and slightly smaller than the European red mite adult. They are white until they feed when they take on the coloration of their prey, usually red or brown.

The eggs are pear shaped, almost transparent, but slightly larger than the round European red mite eggs. The larvae are also transparent and difficult to see without a microscope. Of the five *G. occidentalis* life stages, only the larvae are six legged. All other post-egg stages have eight legs. In all stages, *G. occidentalis* is indistinguishable from *Neoseiulus fallacis* and *Galendromus pyri*, other phytoseiid predatory mites, without a compound microscope.

Habitat (Crops)

Although *G. occidentalis* is a predator of mites in many systems, this page deals solely with orchard trees (apple, plum, peach, cherry, pear). G. occidentalis can thrive in hot, dry climates.

Pests Attacked

Two-spotted spider mites, McDaniel spider mites, yellow spider mites, apple and pear rust mites, Prunus rust mites, blister mites, and European red mites.

Life Cycle

G. occidentalis overwinters as adult mated females in debris under the trees and in crevices on the tree itself. Many die during very cold winters. Emergence begins at first bud, and the mites disperse, looking for either spider mites or apple rust mites. They will also consume European red mites, but they do not seek them out. Eggs are laid on the undersides of leaves, in the opening buds and on the flowers and hatch in 1-4 days. Adults hunt along the leaf midveins during the day, and over the entire leaf surface toward nightfall.

Total development time, from egg to adult, is 6-12 days, depending on the temperature. Females survive about 30 days and lay about 21 eggs. *G. occidentalis* reproduces very quickly, and there may be 8-10 generations per year.

Relative Effectiveness

The presence of apple rust mites ensures a healthy population of predator mites which will be able to control spider mites and European red mites. Growing populations of apple rust mites in May and June encourage the presence of predatory mites, and at moderate numbers, they are not so damaging to the trees as spider mites. Therefore, controlling apple rust mites means keeping predator mite numbers low. (Apple rust mites are generally found on terminal shoots, whereas spider mites are usually on the lower and more central parts of the tree.) If there is about one predator mite per leaf and no more than 25 other mites per leaf, the pest mites should be controlled within 7-10 days.

Pesticide Susceptibility

Pesticide use can not only destroy the predator mite population, but it can destroy less harmful prey species so that the predators will starve. Careful management of *G. occidentalis* and apple rust mites provide the best control of pest mites. If miticides must be used, this should be done before introduction of the predators.

G. occidentalis is highly susceptible to permethrin, methomyl, and esfenvalerate.

Galendromus (=Typhlodromus) pyri (Acarina: Phytoseiidae)

Galendromus pyri is found in orchards throughout the world. It is particularly valuable as a natural enemy because of its ability to feed on many types of food when its primary prey is not available.

Appearance

Adults are pear-shaped and slightly smaller than European red mite adults, which are their primary prey. *G. pyri* adults are white until they feed when they take on the red or brown coloration of their prey. The eggs are pear shaped, almost transparent, but slightly larger than the round

European red mite eggs. The larvae are also transparent and difficult to see without a microscope.

Of the five *G. pyri* life stages, only the larvae are six legged. All other post-egg stages have eight legs. In all stages, *G. pyri* is indistinguishable from *Neoseiulus fallacis* and *Galendromus occidentalis,* other phytoseiid predatory mites, without a compound microscope.

Habitat (Crops)

Many; this page deals only with orchard crops. *G. pyri* thrives in cool, humid climates, especially the orchards of the northeast, northern midwest, and early and late season in northwest orchards.

Pests Consumed

Galendromus pyri prefers European red mite and actively seeks this prey. It will also feed on the two-spotted spider mite and the apple rust mite.

Life Cycle

Mated adult females overwinter in crevices of the tree bark. They emerge on warm spring days, possibly before the tree buds. They live about 20 days, laying approximately 20-30 eggs. Eggs are laid on or in flower buds before flowers open, and afterwards along the ribs of the undersides of leaves. They hatch in 1-3 days, depending on the temperature. The life cycle, which is composed of five stages, egg, larva, protonymph, deutonymph, and adult, is completed three to four times a season. The low numbers in the spring, due to winter mortality, increase as summer progresses.

G. pyri move vigorously over plant surfaces in search of prey. When prey is scarce, they can survive and reproduce on a diet of pollen, fungi and plant fluids.

Relative Effectiveness

Unlike *N. fallacis, G. pyri* does not increase in population as quickly as an expanding pest mite population. However, *G. pyri* survives the winter in greater numbers than *N. fallacis*. In addition, it is able to thrive and will remain in the tree at low prey density, surviving on alternate foods whereas *N. fallacis* will leave. Thus, once the prey population is low, *G. pyri* will prevent its increasing. A mixed population of *G. pyri* and *N. fallacis* is desirable in northeast orchards and northern midwest orchards.

In northwest orchards, *G. pyri* works well with *Galendromus occidentalis*. In early and late season when it is cool, *G. pyri* is more efficient, whereas *G. occidentalis* is more efficient during the hot, dry summers.

Pesticide Susceptibility

Predatory mites are susceptible to many of the chemicals used to combat herbaceous mite infestations. A single application of a chemical considered highly toxic to *G. pyri* at any time during the season, will have a large negative impact on its abundance.

A strain of *G. pyri* that is resistant to pyrethroids has been patented in New Zealand.

The habit of *G. pyri* to overwinter in crevices can be used to advantage in the early spring with a pre-bloom horticultural oil application. This greatly reduces the number of European red mite eggs while not affecting predatory mite populations.

Conservation

It may take up to three years to establish a population of predators high enough to control pest mites. Integrated pest management strategies, as outlined in the tutorial of this guide, can help establish colonies of predatory mites.

Commercial Availability

G. pyri is a common inhabitant of commercial apple orchards in the northeastern United States. In addition, it can be obtained commercially.

Neoseiulus (=Amblyseius) fallacis (Acarina: Phytoseiidae)

This predaceous mite has a strong preference for pest mite species and will travel from tree to tree searching for them. It is found across the continental United States.

Appearance

Adults are pear-shaped and slightly smaller than the European red mite adult. They are white until they feed when they take on the coloration of their prey (usually red or brown). The eggs are pear shaped, almost transparent, but slightly larger than the round European red mite eggs. The larvae are also transparent and difficult to see without a microscope.

Of the five *N. fallacis* life stages, only the larvae are six legged. All other post-egg stages have eight legs. In all stages, *N. fallacis* is indistinguishable from *Galendromus pyri* and *G. occidentalis*, other phytoseiid predatory mites, without a compound microscope.

Habitat (Crops)

Many; this page deals only with fruit trees.

Pests Attacked

In North American orchards, *Neoseiulus fallacis* strongly prefers tetranychid mites—the European red mite and the two-spotted spider mite—and will actively seek these.

Life Cycle

Mated adult females overwinter in crevices of the tree bark if prey are available in the fall. They emerge as early as bloom, but in reduced numbers due to heavy winter mortality. *N. fallacis* increases in number rapidly and adults become numerous by July or August. They live about 20 days and lay an average of 40-60 eggs. Eggs are laid along the ribs of the undersides of leaves. Four to six generations are completed in a season in New York state. *N. fallacis* moves vigorously over plant surfaces in search of prey.

Relative Effectiveness

Because *N. fallacis* is a voracious consumer of mites and because its population increases quickly in relation to its prey, it can overtake an expanding pest population. It develops into the adult stage in about one third the time required by *G. pyri*. However, when *N. fallacis* has reduced the prey population, it will leave the tree in search of more tetranychid mites whereas *G. pyri* thrives on alternate foods. Over the winter, *N. fallacis* has a higher mortality rate than *G. pyri*. Therefore, a mixed population of *N. fallacis* and *G. pyri* is desirable.

Pesticide Susceptibility

Predatory mites are susceptible to many of the chemicals used to combat herbaceous mite infestations. A single application of a chemical considered highly toxic to *N. fallacis* at any time during the season will have a large negative impact on its abundance.

Conservation

The habit of *N. fallacis* to overwinter in crevices can be used to advantage in the early spring with a pre-bloom horticultural oil application. This greatly reduces the number of European red mite eggs while not affecting predatory mite populations.

Pest mite problems are most common where pesticides are heavily used because predatory mite populations are killed along with target species It may take up to three years to establish a population of predators high enough to control pest mites. Integrated pest management strategies, as outlined in the tutorial of this guide, can help establish colonies of predatory mites.

Zetzellia mali (Acarina: Stigmaeidae)

Zetzellia mali is found throughout the United States and exists in most orchards to some extent.

Appearance

Z. mali is smaller and less active than other predator mites. It is bright yellow, although its gut is visibly reddish after feeding. Its eggs are round, bright yellow, and approximately half the size of European red mite eggs. It completes about four life cycles per year, and it does not have the potential to build up as rapidly as the phytoseiid mites. However, it has the ability to persist utilizing various food sources, and so it can eventually reach high numbers.

Habitat (Crops)

Apples orchards.

Pests Attacked

Apple rust mite, European red mite, and two-spotted spider mite.

Life Cycle

Z. mali overwinters as a mated female in crevices and under bark. It prefers to feed on rust mites but will prey on eggs and immature stages of tetranychid mites (European red mite and the two-spotted spider mite) as well as cannibalizing its own. It can survive on pollen, sap and fungal spores, but will not reproduce on this diet. It is slow to explore the tree in search of new prey, so as pest mites move from older leaves to new feeding sites on younger ones, they can escape predation by *Z. mali*.

Relative Effectiveness

Z. mali is not likely to control pest mite infestations alone. It prefers to feed on the apple rust mite and does not eat many prey when it does feed on the tetranychid mites. In addition, it is slow to increase in population and may prey on other predator mite eggs. However, *Z. mali* can be beneficial over the long run in these ways: it can survive on several alternate foods when its preferred prey is not present and thus can live for long periods without prey; because of its feeding versatility, it can eventually reach population densities capable of controlling pest mites; it can complement biological control by other predator mites by feeding on stationary forms of pest mites such as eggs or immature stages (*Neoseiulus fallacis* and *Galendromus pyri* prefer the mobile forms); it can serve as another form of prey for *N. fallacis* and *G. pyri*.

Pesticide Susceptibility

Pest mite problems are most common where pesticides are heavily used because predatory mite populations are killed along with target species. Although *Z. mali* has some resistance to organophosphate insecticides, it is susceptible to endosulfan.

Conservation

Pesticides that are considered moderately toxic to predators may have little or no long term effect on their populations, as a whole, when applied sparingly, although if they are used too often they will have a negative effect. It may take up to three years to establish a population of predators high enough to control pest mites. Integrated pest management strategies, as outlined in the tutorial of this guide, can help establish colonies of predatory mites.

Predatory mites (general)
Euseius tularensis *(Acarina: Phytoseiidae)*

This predatory mite is an important control agent of citrus red mite and citrus thrips in San Joaquin Valley, California citrus orchards. A closely related species, *E. hibisci,* is common in the southern citrus growing region of California. *E. tularensis* prefers to inhabit citrus and *E. hibisci* prefers to inhabit avocados.

Appearance

Adults are pear-shaped and very shiny. They are white when feeding on pollen, yellow when feeding on citrus thrips, and red when they feed on citrus red mite. They avoid direct sunlight and when held on a leaf in the sun they will run rapidly down the main vein or across the leaf.

The eggs are oblong, almost transparent, and slightly larger than the citrus red mite eggs. The six-legged larvae are also transparent. *E. tularensis* is difficult to distinguish from other *Euseius* species without a compound microscope.

Habitat

Citrus

Pests Attacked

Primarily citrus red mite and citrus thrips, however, two-spotted spider mite, immature stages of scale insects and whitefly nymphs are also fed upon. This predatory mite also feeds on pollen and leaf sap.

Life Cycle

E. tularensis overwinters as adults on the sucker shoots in the center of citrus trees. This predatory mite responds to the leaf texture and nutrition and is most abundant when new flush appears in the tree in the spring and fall. New citrus flush is rapidly followed by flowers, petal fall, and fruit development. Thus, the predatory mites are present when citrus thrips are damaging the fruit. Adults hunt along the leaf midveins in the shade in the day, under the calyx of the developing fruit, and over the entire fruit leaf surface toward nightfall. Total development time, from egg to adult, is 6-10 days, at 78-80°F. Females live about 30 days and lay 17 to 27 eggs, depending on the type of food they have available.

Relative Effectiveness

Because this predatory mite is a generalist in its feeding habits, it does not target any particular pest and so it may not regulate their numbers below an economic threshold. Populations of 0.5 to 1.0 per leaf help to reduce citrus red mite and citrus thrips populations.

Pesticide Susceptibility

Various San Joaquin Valley populations of *E. tularensis* have developed resistance to many organophosphate insecticides especially chlorpyrifos. They are still very sensitive to many carbamates (methomyl, formetanate hydrochloride) and pyrethroids used for citrus thrips, caterpillars, and scale control in citrus. Compatible, soft pesticides are sabadilla, spinosad, or spinetoram for citrus thrips control and narrow range petroleum oil, pyriproxyfen, and applaud for California red scale control.

Phytoseiulus persimilis (Acarina: Phytoseiidae)

Phytoseiulus persimilis, a predaceous mite, is one of the mainstays of greenhouse integrated pest management programs for control of spider mites on vegetables and ornamentals in Europe, North America, and elsewhere. This mite was accidentally introduced into Germany from Chili in 1958 and subsequently shipped to other parts of the world, including California and Florida, from Germany.

Appearance

Although extremely small (approximately 0.5 mm or 0.02 inches), *P. persimilis* can be distinguished with a hand lens. It is fast moving, orange to bright reddish orange, has a teardrop-shaped body and long legs, and is slightly larger than its prey. Immatures are a pale salmon color. Eggs are oval, approximately twice as large as the pest mite eggs.

(Note: in the winter, the twospotted spider mite also develops a reddish color, although two dark spots on its abdomen usually distinguish this pest from other mites.)

Habitat

Greenhouses, interior plantscapes, and crops where twospotted spider mites are a problem.

Pests Attacked (Host Range)

This species is a specialized predator of web-spinning spider mites such as the twospotted spider mite. In fact, *P. persimilis* feeds, reproduces, and completes development only on mites in the subfamily Tetranychinae, although it also feeds on young thrips and can be cannabilistic when spider mite prey is unavailable.

Life Cycle

P. persimilis eggs hatch in 2-3 days, and although the larval stage does not feed, the subsequent nymphs and adults feed on all stages of prey. Total time from egg to adult ranges from 25.2 days at 15°C (59°F) to 5.0 days at 30°C (86°F).

The adult female may lay up to 60 eggs during her 50 day-long lifetime at 17-27°C. Generation times of from seven to 17 days are possible, depending on temperature and humidity. Due to its tropical origin, *P. persimilis* does not have a diapause stage and is active year-round in enclosed habitats such as interior plantscapes and greenhouses.

Relative Effectiveness

Adult *P. persimilis* eat from 5-20 prey (eggs or mites) per day, they reproduce more quickly than the spider mites at temperatures above 28°C (82°F), and they feed on all stages of the twospotted spider mite. *P. persimilis* are <u>very</u> voracious. They have the highest consumption rate of all phytoseiids. However, they absolutely must have spider mite prey or they will disperse and/or starve.

Almost 75% of European greenhouse vegetable production relies on *P. persimilis* for spider mite control, and the California strawberry industry uses this mite, along with another beneficial mite, *Neoseiulus* (=*Amblyseius*) *californicus*, to control spider mite infestations in field-grown strawberries. It is also used in interior plantscapes and conservatories. Greenhouse ornamentals growers have long relied on its ability to control spider mites.

Humidity strongly impacts *P. persimilis*' efficacy. Development was observed to almost stop at humidities of 25-30%, and relative humidities below 70% resulted in a reduction in the ability of immatures to molt from one stage to another. In one study, at a relative humidity of 40% (temperature 27°C), only 7.5% of eggs hatched compared to 99.7% at 80% relative humidity (same temperature). Eggs held at a relative humidity of 50% appeared to shrivel at all temperatures from 13-37°C.

Phytoseiid mites use odors (kairomones) associated with mite-infested plants to locate their prey. When *P. persimilis* contacts spider mite webbing, it intensifies its search for prey.

P. persimilis has high dispersal ability and its distribution is highly correlated to that of its prey. However, its ability to disperse is dependent on the environment. If infested plants' leaves touch, dispersal is possible. When the plants have little contact with each other, dispersal is reduced by about 70%. *P. persimilis* moves upward on the plant in search of prey and disperses when prey is scarce. Nymphs do not disperse easily, and are left behind when prey becomes scarce.

Because these mites are such efficient hunters and dispersers, they can cause extinction of their spider mite prey. This is desirable where little or no spider mite damage can be tolerated, such as in ornamental plants. However, in crops where some plant damage is acceptable (e.g., tomatoes and cucumbers), it is desirable to have a stable interaction between predator and prey over an extended period of time.

Typically, *P. persimilis* will eventually exhaust its food supply and starve, and so it must be reintroduced.

Conservation

Relative humidities greater than 60% are required for survival of the predator, particularly through the egg stage.

Harvestmen
Phalangium opilio (Arachnida: Opiliones, Phalangiidae)

HARVESTMAN, DADDY LONGLEGS, HARVEST SPIDER

Phalangium opilio is "the most widespread species of harvestman in the world", occurring natively in Europe, and much of Asia, and having been introduced to North America, North Africa and New Zealand. It is found in a wide range of habitats, including meadows, bogs, forests, and various types of anthropogenic habitats, such as gardens, fields, hedgerows, lawns, quarries, green places in built-up areas, walls and bridges.

Of the many species of harvestmen known, *P. opilio* tends to be the most common in relatively disturbed habitats such as most crops in temperate regions. Like the spiders and most adult mites, harvestmen have two

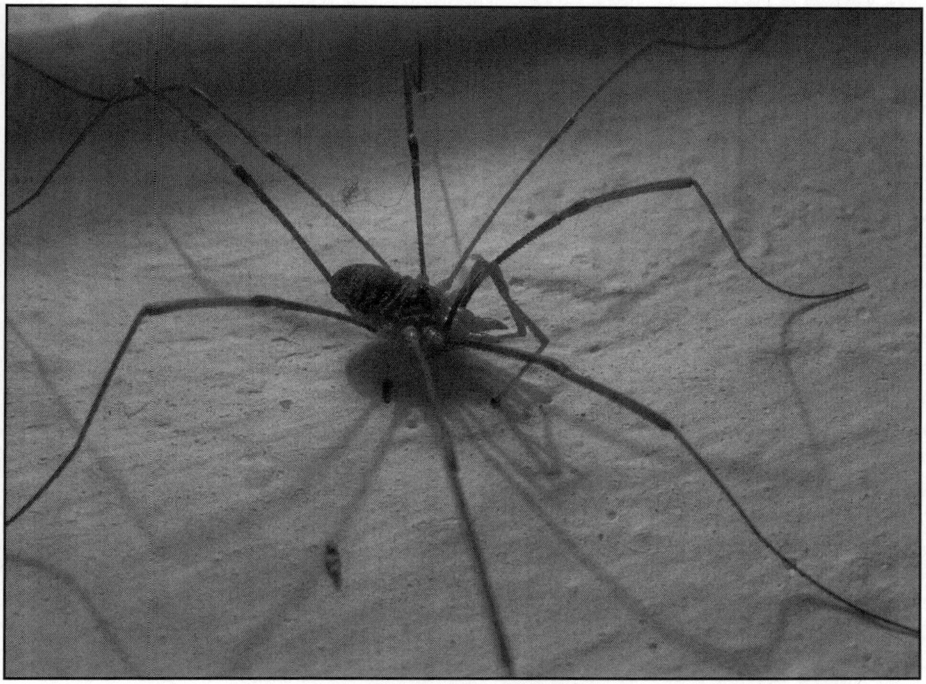

Phalangium opilio

major body sections and eight legs and lack antennae. Unlike spiders, the two body sections of harvestmen are broadly joined and no web spinning organs are present. Harvestmen differ from most mites by their larger size and by having the posterior body section distinctly segmented.

Appearance

The most notable features of *P. opilio* and many other harvestmen are the long, slender legs and short, globular body. Adult body length is approximately 3.5–9 mm, with males generally smaller than females. The upper surface of the body is colored with an indistinct and variable light gray or brown pattern, and the lower surface is typically light cream. Immatures are similar to adults, only smaller and with legs shorter relative to the body size. Eggs are spherical, about 0.4 mm diameter, with a smooth surface and color changing from off-white to dark gray-brown as they mature. They are laid in clusters of around ten to several hundred.

Habitat (Crops)

Harvestmen are often common in crops such as corn, alfalfa, small grains, potatoes, cabbage, strawberries, and apple in most temperate regions of the world.

Pests Attacked

Harvestmen will feed on many soft bodied arthropods in crops, including aphids, caterpillars, leafhoppers, beetle larvae, mites, and small slugs.

Life Cycle

In Europe *P. opilio* has a single generation per year and overwinters as eggs. In parts of North America two or more generations may occur, and eggs, immatures, or adults may overwinter. Eggs are laid in moist areas under rocks, in cracks in the soil, or between the soil and the crowns or recumbent leaves of plants. The eggs hatch in three weeks to five months or more, depending on temperature, and the immatures undergo several molts and reach maturity in two to three months, again depending on temperature.

Relative Effectiveness

Although *P. opilio* by itself appears unable to keep populations of any pest under control, it serves as one member of a complex of generalist predators that exist in many crops and that together are able to help keep pest densities low. In addition to pest arthropods, *P. opilio* also may feed on dead insects and other decaying material, as well as earthworms, other harvestmen, spiders and other beneficial invertebrates. Although its generalist feeding habits and tendency for cannibalism may appear to reduce its value in some situations, they may also allow it to persist in the crop during periods of low pest density and help suppress outbreaks of pests in their early stages.

Pesticide Susceptibility

P. opilio is highly susceptible to at least some broad spectrum insecticides, while some more specific products, such as Bts, appear to be less harmful.

Conservation

Avoid using broad spectrum insecticides as much as possible.

7

Weed Feeders

Due to mankind's ever-increasing mobility, exotic plant species have been extensively introduced to new locations around the world. Often these new species spread rapidly because they have no effective natural enemies in their new locations. In the U.S., it is estimated that 50% to 75% of problem weeds were either accidental or intentional introductions from other areas. For example, purple loosestrife is a valued wildflower in its native Eurasian habitats, existing in small stands and as isolated plants controlled by a complex of natural enemies. It was introduced to North America in the late nineteenth century where it now exists in huge, monotypic stands, choking waterways and crowding out native North American species. Traditional methods of weed control, such as cultivation and the use of herbicides, are not practical or desirable for such wide-ranging stands of weeds, and biological control has become an attractive strategy for control of pest weeds.

Worldwide, classical biological control has successfully controlled introduced weeds on numerous occasions. In North America, there have been successes in controlling nodding thistle (Canada; Kansas, U.S.), ragwort (British Columbia, Canada; California and Oregon, U.S), klamath weed (Ontario, Canada; California, Oregon and southeast Washington, U.S.) alligator weed (Florida, Louisiana, and Texas, U.S.), and water lettuce (Florida). When not completely controlled, pest weeds may be reduced in varying degrees by natural enemy introductions. Sometimes a complex of natural enemies is used in an attempt to control a pest weed.

In addition to introductions of natural enemies in the wild, biological control is used in agriculture to control weeds. Fungal products are available that attack milkweed vine and sicklepod, and Colletotrichum gloeosporioide has been effective in controlling northern jointvetch, a plant pest in rice and soybean crops.

Insects can control weeds by feeding on seeds, flowers, leaves, stems, roots, or combinations of these, or by transmitting plant pathogens, which will infect plants. Other natural enemies of weeds include plant pathogens, nematodes, and vertebrates (fish).

In considering species for introduction to control pest weeds, researchers first determine that the species feeds exclusively on the pest species. Other considerations include the effectiveness of control, favorable host-plant synchronization with the natural enemy's life cycle, the fecundity of the natural enemy, and its heartiness. Researchers have also determined that a successful introduction is more likely if the new locale is similar in ecology and climate to the area of origin, and if trials are made where population densities of the pest plant are similar to those in the new location. The reasons for success or failure are not yet well understood.

Using biological agents for weed control is beneficial because there is less overall expense compared to herbicidal sprays. Once a population of biocontrol agents is established, minimal effort is required to conserve it. Also, the use of natural enemies does not require a high level of technology. Another major benefit is that natural enemies are less disruptive ecologically and so natural biodiversity is maintained.

Some desirable characteristics of weed-feeding natural enemies

- they are specific to one plant species
- they have a negative impact on plant individuals and the population dynamics of the target weed
- they are prolific
- they thrive and become widespread in all habitats and climates that the pest weed occupies
- they are good colonizers

Relative Effectiveness

North American introductions of weed-feeding natural enemies have ranged from very successful, with a 99% reduction of the pest species, to complete failures, with the introduced species unable to become established in the new location.

Predators, parasitoids, or pathogens of the introduced species in the new location may limit its ability to control the pest plant.

ALLIGATORWEED CONTROL AGENT

Agasicles hygrophila (Coleoptera: Chrysomelidae)

ALLIGATORWEED FLEA BEETLE

Agasicles hygrophila is a species of leaf beetle known by the common name **alligator weed flea beetle**. It has been used successfully as an agent of biological pest control against the noxious aquatic plant known as alligator weed (*Alternanthera philoxeroides*).

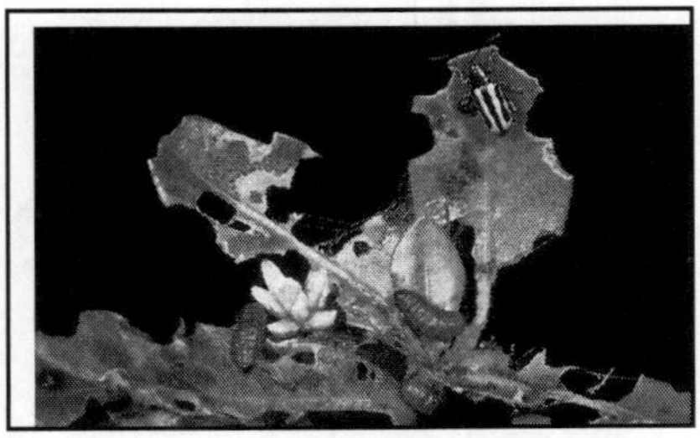

Agasicles hygrophila

This beetle is native to South America but has been imported to areas where alligator weed is a problem. The adult beetle is 5 millimeters in length and black with yellow stripes on its elytra. The female lays approximately 1,000 eggs in her six-week lifetime. Millimeter-long eggs are laid in rows on leaves, and the small yellow larvae emerge and eat the leaves. Adults also feed on the leaves. Defoliation of alligator weed mats kills the weed and clears the infested waterway. This beetle is established in much of the southeastern United States, where it lives on alligator weed only.

Alligatorweed has been a waterway pest in the United States since it was introduced from South America in the 1890s. It forms dense, impenetrable mats that float on the surface and can also exist during drought conditions in a terrestrial form. The flea beetle, *Agasicles hygrophila,* was introduced in 1964 to combat alligatorweed which invades United States waterways from Virginia to southern Florida and along coast waterways to Texas and in California. The success of this program was so remarkable that the U.S. Army Corps of Engineers canceled all herbicide spraying against alligatorweed in Florida in 1968, three years after the beetles were introduced there!

Appearance

Adults are black, about 4-6 mm long, with two longitudinal yellow stripes. The larvae are black, up to 6 mm long when mature third instars; eggs are yellowish, 1.25 mm long and are laid in clusters on the undersides of leaves. Adults have greatly enlarged hind femora for jumping, hence the name flea beetle.

Habitat

Host-range studies in Argentina and Uruguay limited to 14 plant species in 8 families determined that only alligatorweed was suitable as a host plant for *A. hygrophila*. No additional testing was done in North America.

Pests Attacked

Alligatorweed (*Alternanthera philoxeroides*).

Life Cycle

Adults overwinter among the roots and stems of alligatorweed along the margins of waterways. Eggs are laid on the undersides of leaves. Larvae eat the leaves during their development, often leaving transparent windows of uneaten epidermis. Pupation takes place inside the hollow plant stems. The life cycle takes less than one month, and adults live about 48 days.

Relative Effectiveness

The alligatorweed flea beetle is so successful that it is often used as the symbol of biological weed control. Within 4 years, alligatorweed was practically eliminated at the two northern Florida sites where the flea beetle was first introduced. The beetle is less effective in southern Florida and in northern areas of Alabama, Louisiana, Mississippi, and in North and South Carolina. Additional beetles from a colder region in Argentina were released in the Carolinas in 1979, apparently without success.

Pesticide Susceptibility

Susceptible to topically applied toxaphene and methyl parathion, but not to plants grown in water with those pesticides.

DALMATIAN TOADFLAX AND YELLOW TOADFLAX CONTROL AGENTS

Brachypterolus pulicarius (L.) (Coleoptera: Nitidulidae)

Brachypterolus pulicarius was accidentally introduced into North America beginning prior to 1920, presumably as a contaminant of imported ornamental toadflax plants. *B. pulicarius* is now widely distributed in the US and Canada on yellow toadflax, but is apparently less abundant on Dalmatian toadflax.

Appearance

Brachypterolus pulicarius adults are black, oval beetles about 2 mm in length and are found feeding on young toadflax shoots and flower buds. Larvae are 1-4 mm long with short legs and are creamy-white in color; they are found within toadflax flowers and developing fruits.

Habitat

Grasslands, pastures, agricultural fields, and roadsides infested with yellow or Dalmatian toadflax.

Pests attacked

B. pulicarius attacks both yellow and Dalmatian toadflax. Host specificity tests have not been conducted with this insect, so its general host range is unclear. Occasionally, adult beetles may be found on flowers of other species, including dandelion, apple, strawberry, and wild mustards, but no further development is possible on these plants.

Life cycle

Adult beetles emerge from the soil in late spring and are present through mid-summer, feeding on young toadflax shoot tips and, perhaps, on flower buds. After mating, female beetles lay eggs in toadflax flowers.

Young larvae feed on pollen, anthers, and ovaries within the flowers, while older larvae also consume developing seeds. Larvae move freely from flower to flower as they consume floral tissues and seeds. By late summer, mature larvae leave flowers and developing fruits and enter the soil to pupate.

Winter is spent in the pupal stage. There is the one generation per year in the northern US and Canada, though some European reports suggest that two generations are possible.

Relative effectiveness

Flower- and fruit-feeding by *B. pulicarius* larvae may reduce toadflax seed production by more than 75%, but long-term effects on toadflax density are not yet known.

Pesticide susceptibility

Not yet known.

Conservation

The conservation of natural enemies is probably the most important and readily available biological control practice available to growers. Natural enemies occur in all production systems, from the backyard garden to the commercial field. They are adapted to the local environment and to the target pest, and their conservation is generally simple and cost-effective. With relatively little effort the activity of these natural enemies can be observed. Lacewings, lady beetles, hover fly larvae, and parasitized aphid mummies are almost always present in aphid colonies. Fungus-infected adult flies are often common following periods of high humidity. These natural controls are important and need to be conserved and considered when making pest management decisions. In many instances the importance of natural enemies has not been adequately studied or does not become apparent until insecticide use is stopped or reduced. Often the best we can do is to recognize that these factors are present and minimize negative impacts on them. If an insecticide is needed, every effort should be made to use a selective material in a selective manner.

Calophasia lunula (Hufnagel) (Lepidoptera: Noctuidae)

Calophasia lunula is a species of noctuid moth known by the common names toadflax moth and toadflax brocade moth. It is used as an agent of biological pest control against yellow toadflax (*Linaria vulgaris*) and Dalmatian toadflax (*Linaria genistifolia* ssp. *dalmatica*) in areas where they are considered noxious weeds, such as in North America.

The adult is a hairy gray moth about 12 millimeters long. The female lays anywhere from 30 to 80 eggs on the leaves and flowers of toadflax plants. In one to two weeks the larva emerges from the egg and begins feeding. This caterpillar undergoes five molts during a one-month period. The first-stage larva is about 5 millimeters long and dark gray in color. The eyecatching older larvae are black and white spotted with bright yellow stripes. The mature larva is about 4 centimeters long. After this stage it pupates through the winter in a cocoon of chewed plant matter and leaf litter.

Caterpillar

Damage to the plants is done by the voracious feeding of the caterpillars. They consume all the new leaf and flower buds first and then move on to eat stems, leaves, and flowers, often to the point of complete defoliation of large numbers of toadflax plants. This does not necessarily kill the plants but it severely reduces their ability to produce seeds and propagate. Adults feed harmlessly on the plants' nectar and do not contribute to the damage.

This moth is native to Europe. It was first introduced to Canada as a biocontrol agent in the early 1960s, and to the United States shortly thereafter. The moth is now established in many areas. It tends not to establish as easily in colder climates.

Calophasia lunula was introduced in Canada beginning in the early 1960s; US releases were initiated in 1968. Populations are established on yellow toadflax in eastern Canada and on yellow and Dalmatian toadflax in several northwestern US states and British Columbia.

Appearance

C. lunula adults are grey moths with various dark and light wing markings. Moths are about 12-14 mm long, with a wingspan of about 27-30 mm. Moths are inactive during the day, resting on vegetation, but become active at night and feed on nectar from various flowers, including those of toadflax.

Young larvae are 2-3 mm long and pale grey in color. Older larvae have longitudinal yellow, black, and grey stripes with black and white spots, and may reach a length of 35 mm at maturity. Larvae feed on leaves of Dalmatian and yellow toadflax and are active during the day.

Pupae are 14-15 mm long, reddish-brown in color, and contained within white silk cocoons that include soil particles and/or plant debris. Cocoons are usually found on the soil surface or attached to the lower part of plant stems.

Habitat

Grasslands, pastures, agricultural fields, and roadsides infested with yellow or Dalmatian toadflax.

Pests Attacked

C. lunula feeds on both Dalmatian and yellow toadflax. Host specificity appears restricted to species in several genera of the family Scrophulariaceae.

Life Cycle

Adult moths first eclose in late spring and live for several days to more than a week. After mating, females lay eggs individually on toadflax leaves; each female generally lays a total of 30-80 eggs.

Larvae feed on toadflax foliage, at first skeletonizing leaves but later consuming entire leaves. Generally, larvae preferentially feed upon young foliage, but older leaves and flowers are consumed in the absence of young leaves. There are five larval stages. Mature larvae crawl down toadflax stems and construct a silk cocoon into which soil, plant fragments, or frass are added, and then pupate.

Cocooning usually occurs on the soil surface or at the base of a toadflax stem. There are one to three generations per year, depending on weather conditions, with winter spent in the pupal stage.

Relative effectiveness

Calophasia lunula larvae can cause significant defoliation of toadflax plants when populations are large. Though direct plant mortality does not usually occur, defoliated plants generally produce fewer flowers and seeds and may produce fewer nutrients for root storage. However, long-term effects on weed density remain poorly understood.

Pesticide susceptibility

Unknown.

Conservation

The conservation of natural enemies is probably the most important and readily available biological control practice available to growers. Natural enemies occur in all production systems, from the backyard garden to the commercial field. They are adapted to the local environment and to the target pest, and their conservation is generally simple and cost-effective. With relatively little effort the activity of these natural enemies can be observed. Lacewings, lady beetles, hover fly larvae, and parasitized aphid mummies are almost always present in aphid colonies. Fungus-infected adult flies are often common following periods of high humidity. These natural controls are important and need to be conserved and considered when making pest management decisions. In many instances the importance of natural enemies has not been adequately studied or does not become apparent until insecticide use is stopped or reduced. Often the best we can do is to recognize that these factors are present and minimize negative impacts on them. If an insecticide is needed, every effort should be made to use a selective material in a selective manner.

Eteobalea intermediella Riedl (Lepidoptera: Cosmopterygidae)

Eteobalea intermediella is a moth of the Cosmopterigidae family. It is native to an area ranging from Central and Southern Europe, east to the Caucasus, Asia Minor, the Near and Middle East towards Central Asia. It is also present in North Africa (Morocco, Algeria and Tunisia). It has been introduced to the United States and Canada.

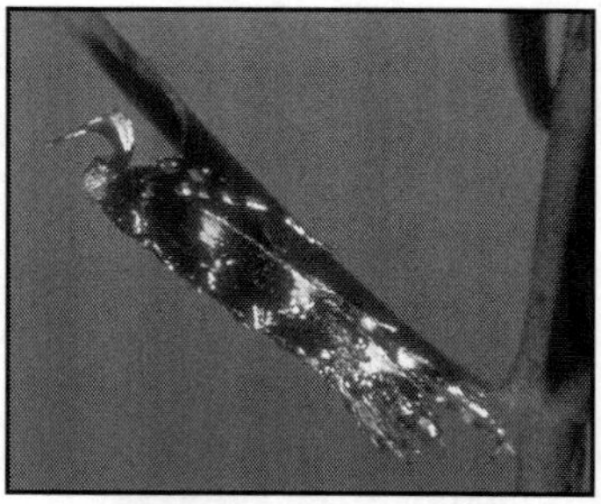

Eteobalea intermediella

The wingspan is 16–18 mm. Adults are on wing from late April to early November in one to two generations.

The larvae feed on the roots of *Linaria vulgaris, Linaria genistifolia, Linaria pontica, Linaria dalmatica* and *Anarrhinum bellidifolium*.

Eteobalea intermediella was approved for release in the US in 1995. A few field releases have been made in western Canada and the western US, but no established populations have yet been confirmed.

Appearance

Eteobalea intermediella adults are slender moths about 8-9 mm long, with a wingspan of 16-18 mm. Wings are dark brown with black and white spots. Adults are generally active in the evening and before dawn.

Larvae are cream-colored caterpillars with brown head capsules, and reach a maximum length of about 12 mm. They are found in tunnels within toadflax roots.

E. intermediella adults and larvae are very similar to those of *E. serratella*; the two species may be separated only by examining male genitalia or the egg chorion.

Habitat

Grasslands, pastures, agricultural fields, and roadsides infested with yellow or Dalmatian toadflax.

Pests attacked

Eteobalea intermediella attacks both Dalmatian and yellow toadflax. Host specificity tests indicate that this species utilizes a small number of perennial Eurasian species in the family Scrophulariaceae; no North American plants so far tested were attacked.

Life cycle

Adults appear in late spring and again in late summer or early fall. Moths live for several weeks but apparently do not feed. After mating, females lay eggs, in loose strings of 3-8, in leaf axils on the lower portion of a toadflax stem. Each female may lay up to 180 eggs. Newly-hatched larvae enter the stem and mine down into toadflax roots.

Larvae feed within the root cortex, constructing silk-lined tunnels. Generally, 3-7 larvae may be found within the root system of a toadflax plant, depending on the size of the roots. Mature larvae tunnel back to the root crown area or up into the base of a stem and construct a chamber in which pupation occurs.

There are two, and possibly three, generations per year in Europe. Winter is spent as larvae within toadflax roots.

Relative effectiveness

Larval root boring by *Eteobalea* spp. does not usually kill toadflax plants directly, but does decrease flowering and seed production and reduce plant vigor. There is not yet sufficient information available to predict the efficacy of this agent under North American field conditions.

Pesticide susceptibility

Unknown.

Conservation

The conservation of natural enemies is probably the most important and readily available biological control practice available to growers. Natural enemies occur in all production systems, from the backyard garden to the commercial field. They are adapted to the local environment and to the target pest, and their conservation is generally simple and cost-effective. With relatively little effort the activity of these natural enemies can be observed. Lacewings, lady beetles, hover fly larvae, and parasitized aphid mummies are almost always present in aphid colonies. Fungus-infected adult flies are often common following periods of high humidity. These natural controls are important and need to be conserved and considered when making pest management decisions. In many instances the importance of natural enemies has not been adequately studied or does not become apparent until insecticide use is stopped or reduced. Often the best we can do is to recognize that these factors are present and minimize negative impacts on them. If an insecticide is needed, every effort should be made to use a selective material in a selective manner.

Eteobalea serratella Treitschke (Lepidoptera: Cosmopterygidae)

Eteobalea serratella is a moth in the Cosmopterigidae family. It is found in most of Europe, except the Benelux, Great Britain, Ireland, Iceland, Fennoscandia and the Baltic States. It was approved for release in the United States in 1995 for the biological control of toadflax. A few field releases have been made in western Canada and the western United States, but no established populations have been confirmed.

The wingspan is 16–18 mm. Adults are dark brown with white and copper spots. There is one generation per year.

The larvae feed on *Linaria vulgaris*. They are off-white with brown heads. They develop through five instars and reach a length of up to 12 mm. First instar larvae bore into the root crown where they feed on tissue inside tunnels they carve and line with silk. Mature larvae overwinter in roots and then pupate in spring inside coccoons within the root crown.

Eteobalea serratella was approved for release in the US in 1995. A few field releases have ben made in western Canada and the western US, but no established populations have yet been confirmed.

Appearance

Eteobalea serratella adults are slender moths about 8-9 mm long, with a wingspan of 16-18 mm. Wings are dark brown with black and white spots. Adults are generally active in the evening and before dawn.

Larvae are cream-colored caterpillars with brown head capsules, and reach a maximum length of about 12 mm. They are found in tunnels within toadflax roots.

E. serratella adults and larvae are very similar to those of *E. intermediella*; the two species may be separated only by examining male genitalia or the egg chorion.

Habitat

Grasslands, pastures, agricultural fields, and roadsides infested with yellow or Dalmatian toadflax.

Pests attacked

Eteobalea serratella appears to primarily utilize yellow toadflax, with Dalmatian toadflax as a relatively poor host. Host specificity tests indicate that this species utilizes a small number of perennial Eurasian species in the family Scrophulariaceae; no North American plants so far tested were attacked.

Life cycle

Adults appear in early summer, living for several weeks but apparently not feeding. After mating, females lay eggs, in loose strings of 3-8, at or near the base of a toadflax stem. Each female may lay up to 180 eggs.

Newly-hatched larvae enter the stem and mine down into toadflax roots. Larvae feed within the root cortex, constructing silk-lined tunnels. Generally, 3-7 larvae may be found within the root system of a toadflax plant, depending on the size of the roots. Mature larvae tunnel back to the root crown area or up into the base of a stem and construct a chamber in which pupation occurs.

There is one generation per year in Europe, with winter spent in the larval stage within toadflax roots.

Relative effectiveness

Larval root boring by *Eteobalea* spp. does not usually kill toadflax plants directly, but does decrease flowering and seed production and reduce plant vigor. There is not yet sufficient information available to predict the efficacy of this agent under North American field conditions.

Pesticide susceptibility

Unknown.

Conservation

The conservation of natural enemies is probably the most important and readily available biological control practice available to growers. Natural enemies occur in all production systems, from the backyard garden to the commercial field. They are adapted to the local environment and to the target pest, and their conservation is generally simple and cost-effective. With relatively little effort the activity of these natural enemies can be observed. Lacewings, lady beetles, hover fly larvae, and parasitized aphid mummies are almost always present in aphid colonies. Fungus-infected adult flies are often common following periods of high humidity. These natural controls are important and need to be conserved and considered when making pest management decisions. In many instances the importance of natural enemies has not been adequately studied or does not become apparent until insecticide use is stopped or reduced. Often the best we can do is to recognize that these factors are present and minimize negative impacts on them. If an insecticide is needed, every effort should be made to use a selective material in a selective manner.

Gymnetron antirrhini Paykull (Coleoptera: Curculionidae)

Gymnetron antirrhini was accidentally introduced into North America after 1900, presumably as a contaminant of imported ornamental toadflax plants. *G. antirrhini* now appears to occur wherever yellow toadflax is found in the US and Canada, but is apparently much less abundant on Dalmatian toadflax. A "strain" of *G. antirrhini* presumably more host-specific to Dalmatian toadflax was approved for release in the US in 1995. (Background information about Dalmation toadflax and yellow toadflax is available.)

Appearance

Gymnetron antirrhini adults are black, oval beetles about 2 mm long, with a pronounced snout. They may be found feeding on toadflax stems, buds, flowers, or fruits.

Larvae are 2-4 mm long, legless, and C-shaped when viewed from the side. They are creamy-white in color with a tan head capsule. Larvae are found within toadflax seed capsules.

Habitat

Grasslands, pastures, agricultural fields, and roadsides infested with yellow or Dalmatian toadflax.

Pests attacked

G. antirrhini feeds on yellow and Dalmatian toadflax; different "strains" of this insect may preferentially utilize individual toadflax species. In general, *G. antirrhini* appears to only utilize several Eurasian Linaria spp. as hosts.

Life cycle

Adult weevils emerge in late spring or early summer, and feed primarily on young toadflax shoots and buds. After mating, female beetles lay eggs inside toadflax flowers, inserting them singly through the ovary wall. Oviposition causes abnormal growths within the ovary, and ovules and young seeds adjacent to these growths become distended and yellowish.

Larvae hatch from eggs and feed on these "inactivated," abnormal seeds. Larval development consists of three stages and last about three to five weeks. Mature larvae construct cells within developing seed capsules and then pupate.

In several weeks, adults eclose and may feed on toadflax shoots for a short period of time before entering diapause. Winter is spent as an adult in plant litter or within old toadflax seed capsules. There is one generation per year.

Relative effectiveness

G. antirrhini does not kill toadflax plants, but larval seed feeding can reduce a plant's seed production by more than 80%.

Pesticide susceptibility

Unknown.

Conservation

The conservation of natural enemies is probably the most important and readily available biological control practice available to growers. Natural enemies occur in all production systems, from the backyard garden to the commercial field. They are adapted to the local environment and to the target pest, and their conservation is generally simple and cost-effective. With relatively little effort the activity of these natural enemies can be observed. Lacewings, lady beetles, hover fly larvae, and parasitized aphid mummies are almost always present in aphid colonies. Fungus-infected adult flies are often common following periods of high humidity. These natural controls are important and need to be conserved and considered when making pest management decisions. In many instances the importance of natural enemies has not been adequately studied or does not become apparent until insecticide use is stopped or reduced. Often the best we can do is to recognize that these factors are present and minimize negative impacts on them. If an insecticide is needed, every effort should be made to use a selective material in a selective manner.

Gymnetron linariae Panzer (Coleoptera: Curculionidae)

Gymnetron linariae was approved for release in the US in 1995. Several releases have been made in western Canada and the US; small populations may be established in Canada, but no confirmed establishment has yet been reported in the US.

Appearance

Gymnetron linariae adults are small, black, oval weevils about 2 mm long, with a noticeable snout. They may be found feeding on toadflax shoots.

Larvae are small, legless, and C-shaped when viewed laterally; presumably, they are creamy-white in color with a yellowish head capsule as are larvae of other *Gymnetron* spp. Larvae are found within galls on toadflax roots.

Habitat

Grasslands, pastures, agricultural fields, and roadsides infested with yellow or Dalmatian toadflax.

Gymnetron linariae

Pests attacked

Gymnetron linariae attacks both Dalmatian and yellow toadflax. Based on host specificity tests, only a few Eurasian Linaria spp. are suitable hosts for this weevil.

Life cycle

Gymnetron linariae adults appear in late spring, and apparently live for up to several months. After several weeks of feeding on young toadflax shoots, adults mate and females begin laying eggs in toadflax roots. Eggs are deposited singly into small pockets chewed into the root tissue by the female weevil and covered with excrement. Most eggs are laid in the root crown area, within several inches of the soil surface.

In response to female oviposition and, perhaps, larval feeding, round or oval galls are formed on toadflax roots within which *G. linariae* larvae feed and develop. Up to 200 galls may be found on a single toadflax plant; individual galls may fuse to form "gall masses" consisting of 40 or more galls. Larval development consists of three stages and takes two or three months.

Pupation occurs within galls in late summer. Some newly-eclosed adults may remain within galls and enter diapause, while others may emerge and feed on toadflax shoots for a week or two before returning to the soil to overwinter.

There is one generation per year.

Relative effectiveness

The potential impact of *Gymnetron linariae* on toadflax plants is not yet known. Presumably, root galls will reduce plant vigor by acting as a nutrient "sink," as has been suggested for other plant-galling insects.

Pesticide susceptibility

Unknown.

Conservation

The conservation of natural enemies is probably the most important and readily available biological control practice available to growers. Natural enemies occur in all production systems, from the backyard garden to the commercial field. They are adapted to the local environment and to the target pest, and their conservation is generally simple and cost-effective. With relatively little effort the activity of these natural enemies can be observed. Lacewings, lady beetles, hover fly larvae, and parasitized aphid mummies are almost always present in aphid colonies. Fungus-infected adult flies are often common following periods of high humidity. These natural controls are important and need to be conserved and considered when making pest management decisions. In many instances the importance of natural enemies has not been adequately studied or does not become apparent until insecticide use is stopped or reduced. Often the best we can do is to recognize that these factors are present and minimize negative impacts on them. If an insecticide is needed, every effort should be made to use a selective material in a selective manner.

Gymnetron netum (Germar) (Coleoptera: Curculionidae)

Gymnetron netum was accidentally introduced into North America in the early 1900's, presumably as a contaminant of imported ornamental toadflax plants. *G. netum* has been reported from a number of states across the northern US and from British Columbia. Most collections were reported from yellow toadflax, though this species has also been collected from Dalmatian toadflax in the northwestern US. A presumed Dalmatian toadflax-specific "strain" of *G. netum* is under investigation but has not yet been approved for release in the US.

Appearance

Gymnetron netum adults are oval, dark grey beetles about 2-3 mm long that have a noticeable snout. They may be found on young toadflax shoots. *G. netum* larvae are 2-4 mm long, legless, and C-shaped when viewed from the side. They are creamy-white in color with a light brown head capsule. Larvae are found within toadflax seed capsules.

Habitat

Grasslands, pastures, agricultural fields, and roadsides infested with yellow or Dalmatian toadflax.

Pests attacked

Gymnetron netum attacks both yellow and Dalmatian toadflax; there may be "strains" of this insect that preferentially attack a single toadflax species. Studies examining the host specificity of this species are currently in progress.

Life cycle

The life cycle of *G. netum* appears to be quite similar to that of *G. antirrhini*. Adult weevils emerge in late spring or early summer, and feed on young toadflax shoots and buds for a week or two. After mating, female beetles lay eggs inside toadflax flowers, inserting them singly through the ovary wall. Females lay a total of about 20-60 eggs. Oviposition causes abnormal growths within the ovary, and ovules and young seeds adjacent to these growths become distended and yellowish.

Larvae hatch from eggs and feed on these "inactivated," abnormal seeds. Larval development consists of three stages and last about three to five weeks. Mature larvae construct cells within developing seed capsules and then pupate. In several weeks, adults eclose and may feed on toadflax shoots for a short period of time before entering diapause. Winter is spent as an adult in plant litter or within old toadflax seed capsules.

There is one generation per year.

Relative effectiveness

Gymnetron netum feeds on developing seeds and will, presumably, reduce seed production on attacked plants, as documented for *G. antirrhini*. This agent will not directly kill toadflax plants, and long-term effects on toadflax abundance are not yet known. *Gymnetron netum* may be displaced by populations of *G. antirrhini* when the two species occur together.

Pesticide susceptibility

Unknown.

Conservation

Specifics are unknown. The conservation of natural enemies is probably the most important and readily available biological control practice available to growers. Natural enemies occur in all production systems, from the backyard garden to the commercial field. They are adapted to the local environment and to the target pest, and their conservation is generally simple and cost-effective. With relatively little effort the activity of these natural enemies can be observed. Lacewings, lady beetles, hover fly larvae, and parasitized aphid mummies are almost always present in aphid colonies. Fungus-infected adult flies are often common following periods of high humidity. These natural controls are important and need to be conserved and considered when making pest management decisions. In many instances the importance of natural enemies has not been adequately studied or does not become apparent until insecticide use is stopped or reduced. Often the best we can do is to recognize that these factors are present and minimize negative impacts on them. If an insecticide is needed, every effort should be made to use a selective material in a selective manner.

Mecinus janthinus Germar (Coleoptera: Curculionidae)

Mecinus janthinus was approved for release in the US in 1995. Field releases have been made at sites in western Canada and the western US. Some Canadian populations are established , but establishment at US sites has not yet been confirmed.

Appearance

Mecinus janthinus adults are shiny black, slender weevils about 3-4 mm long, with a pronounced snout. Beetles are found on toadflax foliage. The legless larvae are creamy-white in color with a light-brown head capsule, and appear C-shaped when viewed from the side. *M. janthinus* larvae are found tunneling within toadflax stems.

Mecinus janthinus

Habitat

Grasslands, pastures, agricultural fields, and roadsides infested with yellow or Dalmatian toadflax.

Pests attacked

M. janthinus attacks both Dalmatian and yellow toadflax. Host specificity tests indicate that only a few plants within the genus *Linaria* are utilized as hosts.

Life cycle

Mecinus janthinus adults emerge in late spring or early summer and feed on toadflax leaves and stems, and may live for several weeks to about a month. After a feeding period, adults mate and female weevils begin laying eggs in toadflax stems. Eggs are deposited singly in cavities chewed into the stems and sealed with a lid of chewed plant material.

Young larvae begin feeding within toadflax stems, creating tunnels that increase in size and length as larvae mature. Wilted shoots are an external symptom of this larval feeding. Larval development is completed in about a month, and mature larvae construct a cell within the mined stem in which pupation occurs.

Pupal development is completed in several weeks; newly-eclosed adults remain within pupal cells and overwinter. There is one generation per year.

Relative effectiveness

Mecinus janthinus larvae mine in toadflax stems. Though this damage does not typically kill plants, mined stems wilt and die with a corresponding reduction in flowering and seed production. This impact may be more pronounced in plants experiencing drought stress. However, long-term effects of *M. janthinus* on weed density are not yet known.

Pesticide susceptibility

Unknown.

Conservation

Specifics are unknown. The conservation of natural enemies is probably the most important and readily available biological control practice available to growers. Natural enemies occur in all production systems, from the backyard garden to the commercial field. They are adapted to the local environment and to the target pest, and their conservation is generally simple and cost-effective. With relatively little effort the activity of these natural enemies can be observed. Lacewings, lady beetles, hover fly larvae, and parasitized aphid mummies are almost always present in aphid colonies. Fungus-infected adult flies are often common following periods of high humidity. These natural controls are important and need to be conserved and considered when making pest management decisions. In many instances the importance of natural enemies has not been adequately studied or does not become apparent until insecticide use is stopped or reduced. Often the best we can do is to recognize that these factors are present and minimize negative impacts on them. If an insecticide is needed, every effort should be made to use a selective material in a selective manner.

DIFFUSE KNAPWEED AND SPOTTED KNAPWEED CONTROL AGENTS

Agapeta zoegana (Lepidoptera: Cochylidae)

Agapeta zoegana is a species of moth known as the sulphur knapweed moth and the yellow-winged knapweed root moth. It is used as an agent of biological pest control against noxious knapweeds, particularly spotted knapweed (*Centaurea maculosa*) and diffuse knapweed (*Centaurea diffusa*).

The adult moth is bright yellow with areas of brown on its wings. It is about 11 millimeters long. The adult lives for only a few days, during which time the female lays eggs on stems and leaves. The larva, a root miner, moves to the root of the plant and tunnels through the cortex as it feeds. This causes moderate damage to the plant.

This moth is native to Eurasia. It was introduced to the western and west-central United States starting in the 1980s and it is now established in knapweed populations. It reduces the spread of the plants in areas where mechanical and chemical control is not practical. It lives on spotted and diffuse knapweed and has not been found to damage other plants.

Flight is from May to August from dusk, and is attracted to light.

Agapeta zoegana

Agapeta zoegana, from Europe, was first released in the United States in 1984. Its release is part of a program to introduce a complex of spotted and diffuse knapweed enemies to help control these natives of Eastern Europe and Asia. In North America spotted and diffuse knapweed are efficient competitors and since being introduced around the turn of the century have spread throughout the northern tier of states and as far south as Nebraska and Virginia. They favor disturbed sites and overgrazed rangelands and cause a reduction of forage for cattle of up to 100% in infested areas. Spotted knapweed and diffuse knapweed also prosper on grasslands and riparian sites and displace native vegetation in these habitats.

Although knapweed can be controlled by extensive cultivation or herbicide application, its existence on stream banks and rough terrain and the typically expansive areas of infestation make these methods impractical and/or economically unfeasible. Therefore, an integrated pest management approach is being sought to help in controlling this weed.

A. zoegana has been released in Arizona, California, Colorado, Idaho, Minnesota, Montana, Nebraska, Nevada, North Dakota, Oregon, South Dakota, Utah, Washington, Wisconsin, and Wyoming. The moth has established in these states.

Appearance

A. zoegana is a small bright yellow moth, 10 mm in length, with brown wing bands. The adults may be found resting vertically on the knapweed stems or under the leaves. They have the appearance of dead or dying knapweed leaves.

Habitat

Spotted and diffuse knapweed-infested areas.

Pests Attacked

A. zoegana is considered host-specific to spotted and diffuse knapweed. It has been tested on fifty-one plant species including closely related native species and species of economic importance such as safflower.

Life Cycle

Adult moths emerge from knapweed roots in early July through early September. Mating takes place within twenty-four hours after emergence and the mated female begins laying eggs the next day. The eggs are laid in the stem crevices and on the leaves of the spotted and diffuse knapweed plants. The larvae hatch from the eggs in seven to ten days and move immediately to the root crown and mine into the root.

Larvae are capable of killing small rosettes and then moving a few centimeters to another knapweed plant to feed. The larvae pupate in the root in midsummer. Adults live from eleven to fourteen days with each female laying from twenty-one to seventy-eight eggs in her lifetime. The moths are strong fliers and will invade new knapweed patches.

Pesticide susceptibility

Not yet known.

Commercial availability

In some states, *A. zoegana* adults may be obtained at no cost from state weed management agencies. Several commercial suppliers can also provide *A. zoegana* adults.

Relative Effectiveness

It is anticipated that this moth will be an effective control agent in conjunction with other root boring and seed feeding insects. Spotted and diffuse knapweed patches are hard to find in Europe because they are kept under control with the native insects and diseases.

Conservation

Sites that are chosen for *A. zoegana* releases should be considered for long term availability. It is wise to consider at least a time commitment of ten years without disturbance of the site by development or pesticide use. For best establishment results *A. zoegana* should be released in areas where there are green knapweed rosettes in August and in the area of the knapweed patch where the knapweed is abundant but not a monoculture.

Bangasternus fausti (Coleoptera: Curculionidae)

Bangasternus fausti is a species of true weevil known as the broad-nosed seed head weevil. It is used as an agent of biological pest control against noxious knapweeds, particularly spotted knapweed (*Centaurea maculosa*), squarrose knapweed (*Centaurea virgata* ssp. *squarrosa*), and diffuse knapweed (*Centaurea diffusa*).

The adult weevil is dark gray and hairy and about 4 millimeters long. The female lays eggs on the flower heads at an early stage of development and stem tips. Upon emergence the larva burrows into the flower head or makes its way there by tunneling through the stem. It feeds upon the developing seeds, often consuming them entirely. If any other insects invade the flower head, the larva attacks them. It pupates inside the seed head.

This weevil is native to southern Europe and the Middle East. It was first released as a knapweed biocontrol in the 1980s in Oregon, and it is currently established in the Pacific Northwest. It is host-specific to invasive knapweeds and has not been known to attack any native plants

Bangasternus fausti (Reitter), is a native weevil of Europe and was released in the United States in 1991. The weevil has been released in California, Colorado, Idaho, Minnesota, Montana (established), Nebraska (established), Oregon (established), South Dakota, Washington, and Wyoming as part of a program to control spotted and diffuse knapweed (*Centaurea maculosa* Lam. and *C. diffusa*). The states listed without confirmed establishment reflect 1994 and 1995 *B. fausti* releases.

Appearance

B. fausti is a small grey brown weevil (4 mm in length) with a blunt snout that appears to have been cut off. *B. fausti* may be found feeding on spotted or diffuse knapweed foliage in the spring and in the summer on the flowers.

Habitat

Spotted and diffuse knapweed are weed species that can be found throughout the northern tier of states and as far south as Nebraska and Virginia. These highly competitive weed species favor and establish quickly on disturbed sites and overgrazed rangeland. Both weeds will invade well established grassland communities and out compete the native vegetation. The release of *B. fausti* is part of a program to introduce a complex of spotted and diffuse knapweed enemies to help control these weeds.

Pests Attacked

B. fausti is considered host specific and limited to a few species in the genus *Centaurea*, specifically *C. diffusa* (diffuse knapweed), *C. maculosa* (spotted knapweed), *C. calcitrapa* (purple star-thistle), *C. squarrosa* (squarrose knapweed), and will rarely attack *C. solstitialis* (yellow star- thistle), all of which are introduced weeds. Extensive host specificity tests were conducted in the field and laboratory in Europe before *B. fausti* was cleared and released in the United States.

Life Cycle

Adult weevils emerge from the seedheads and ground litter in May and early June. Egg laying begins within one to two weeks after emergence and continues through mid-August. The eggs hatch in eight to twelve days and the larvae then mine into the knapweed seedhead and begin feeding. It takes approximately thirty-two days for the weevil to develop from egg to the adult stage.

Pesticide Susceptibility

Not yet known.

Effectiveness

B. fausti larvae destroy up to one hundred percent of the seeds in a knapweed seedhead and any seeds that develop do not germinate. This weevil is part of a complex of insect biocontrol agents that contribute to the seed reduction in spotted and diffuse knapweed.

Chaetorellia acrolophi (Diptera: Tephritidae)

Chaetorellia acrolophi is a species of tephritid fruit fly known as the knapweed peacock fly. It is used as an agent of biological pest control against noxious knapweeds, especially spotted knapweed (*Centaurea maculosa*).

The adult fly is light greenish brown with brown-banded wings and iridescent green eyes. It is 3 to 5 millimeters long. The female lays about 70 eggs beneath the bracts on immature flower heads. When the larva emerges from the egg a few days later it burrows into the flower head and feeds on the developing florets. As the larva grows it begins to feed on the developing seeds, often consuming the entire contents of the flower head during its two-week larval stage. It pupates inside the empty flower head.

This fly is native to Europe. It was first released as a knapweed biocontrol in the 1990s in Montana, and it is currently established in much of the western United States. It is host-specific to invasive knapweeds, preferring spotted knapweed, and has not been known to attack any native plants. The fly is often outcompeted by other species that attack knapweed flower heads, such as certain weevils, so it does best where there are no competitors.

Chaetorellia acrolophi, a seedhead fly, is a native of Europe and was cleared for release in the United States in 1992. The fly has been released in Montana, Oregon, Washington and Wyoming. Recovery has been confirmed in Oregon. Releases are limited by the availability of this biocontrol agent in Europe. This fly is being released to enhance the complex of biocontrol insects for spotted knapweed control.

Appearance

C. acrolophi is a small (3 mm) yellow-brown fly with bright green eyes and light brown wing bars. This fly is found resting or displaying on the buds of spotted knapweed in the early summer.

Habitat

Spotted and diffuse knapweed are weed species that can be found throughout the northern tier of states and as far south as Nebraska and Virginia. These highly competitive plants favor and establish quickly on disturbed sites and overgrazed rangeland. Both weeds will invade well established grassland communities and out-compete the native vegetation. The release of *C. acrolophi* is part of a program to introduce a complex of spotted and diffuse knapweed enemies to help control these weeds.

Pests Attacked

C. acrolophi has a limited host range with a preference for species in the *Centaurea* subgenus *Acrolophus*. *C. acrolophi* was tested against fifty-eight plant species in the family *Asteraceae* including United States endangered and economically important species.

Life Cycle

Adults begin to emerge from the spotted knapweed seedheads in the spring from mid May into early June. Mating begins as soon as the females emerge and eggs are laid within a few days of mating. The eggs hatch in four to five days, then the larvae burrow into the knapweed flower bud and start feeding. The overwintering generation larvae pupate in the spring. In her lifetime, the female fly will lay an average of sixty-nine eggs. She may lay as many as twelve eggs each day. The adults can live up to four weeks.

Pesticide Susceptibility

Not yet known.

Relative Effectiveness

C. acrolophi larvae considerably reduce the seed production of spotted knapweed, typically destroying the entire contents of an infested seedhead. The effect on diffuse knapweed is not clear at this time. The larvae can develop on diffuse knapweed but in field tests *C. acrolophi* did not lay eggs on this plant.

Conservation

C. acrolophi release sites should be selected considering long term availability. Ten years without the threat of being disturbed by development or pesticide use is required.

Cyphocleonus achates (Coleoptera: Curculionidae)

Cyphocleonus achates is a species of true weevil known as the **knapweed root weevil**. It is native to southern Europe and the Mediterranean and is used as an agent of biological pest control against noxious knapweeds, especially spotted knapweed (*Centaurea maculosa*).

The root boring weevil, *Cyphocleonus achates,* was first released in the United States in 1987. Its release is part of a program to introduce a complex of spotted and diffuse knapweed enemies to help control these natives of Eastern Europe and Asia. In North America spotted and diffuse knapweed are efficient competitors and since being introduced around the turn of the century have spread throughout the northern tier of states and as far south as Nebraska and Virginia. They favor disturbed sites and overgrazed rangelands and cause a reduction of forage for cattle of up to 100% in infested areas. Spotted knapweed and diffuse knapweed also prosper on grasslands and riparian sites and displace native vegetation in these habitats.

Although knapweed can be controlled by extensive cultivation or herbicide application, its existence on stream banks and rough terrain and the typically expansive areas of infestation make these methods impractical and/or economically unfeasible. Therefore, an integrated pest management approach is being sought to help in controlling this weed.

C. achates

C. achates has been released in Arizona, California, Colorado, Idaho, Minnesota, Montana, Nebraska, Nevada, Oregon, South Dakota, Utah, Washington, and Wyoming as part of a program to control spotted and diffuse knapweed. Populations are established in Colorado, Montana, and Wyoming, and individuals have been recovered in Oregon. It has also become established in Minnesota (N. Northrop, personal comm.).

Appearance

C. achates is a large, 20 mm long, brown-gray mottled weevil. Its markings and coloration blend with the ground and the mature seedhead on the knapweed plant. Its habit is to sit perfectly still and when disturbed to drop to the ground and play dead. *C. achates* does not fly.

Habitat

Spotted and diffuse knapweed-infested areas.

Pests Attacked

C. achates is considered host specific to spotted and diffuse knapweed. The host tests were conducted on seventy-one plant species, and development from larva to adult occurred on four of the seventy-one plant species: *Centaurea maculosa* (European and American bio-types), *C. diffusa* (European and American bio-types), *C. micranthos* (European), and *C. vallesiaca* (European), all of which belong to the subgenus *Acrolophus*.

Life Cycle

C. achates overwinters as larvae in the tap root of spotted and diffuse knapweed. Adults emerge from late July through late September. The adults feed on the knapweed leaves, preferring tender leaves in the center of the rosettes. The eggs are laid on the plant root just below the soil surface and hatch in ten to twelve days. The larvae burrow into the root and commence feeding on the root of the spotted and diffuse knapweed plant. Multiple larvae may be in the same root. The female weevil copulates many times during the egg laying season and lays an average of sixty-five eggs during her lifetime. *C. achates* produce one generation per year.

The adult weevil is dark gray and white speckled and 14 to 15 millimeters long. The female lays about 100 eggs during her 8- to 15-week adult lifespan. The eggs are deposited one by one in notches the female digs into the root crown of the knapweed. The larva emerges from the egg and tunnels into the root cortex where it feeds on the plant tissue. Sometimes a gall is produced in the feeding area. Adults feed on the leaves but it is the larva that does the most damage to the plant. Small plants can be killed by the larva's destruction of the root tissue. In their native range, most C. achates overwinter as 2nd instars, though in the introduced range, the vast majority overwinter as 1st instars.

Pesticide susceptibility

Not yet known.

Effectiveness

Plants are stunted and survive one season after being infested with *C. achates*. Larvae destroy the interior of the tap root and expose it to bacterial and fungal invasion which helps kill the plant.

It is anticipated that this weevil will be an effective control agent in conjunction with other root boring and seed feeding insects. Spotted and diffuse knapweed patches are hard to find in Europe because they are kept under control with the native insects and diseases.

Conservation

Long-term availability is important in sites that are chosen for *C. achates* releases. It is wise to consider sites that will be undisturbed by development or pesticide use for at least ten years. The weevil needs high soil temperature to complete development and cannot establish in areas with a short growing season. Therefore this weevil should be released in areas or microclimates where the soil is partially exposed and absorbs heat. The weevils do best on a south facing slope that is somewhat barren with rocky exposure.

Larinus minutus (Coleoptera: Curculionidae)

Larinus minutus is a species of true weevil known as the lesser knapweed flower weevil. It is used as an agent of biological pest control against noxious knapweeds, especially diffuse knapweed (*Centaurea diffusa*) and spotted knapweed.

The adult weevil is dark mottled brown with a long snout. It is 4 or 5 millimeters long in total. It is active throughout the summer with a 14-week maximum adult lifespan. During this time the female lays up to 130 eggs, depositing them in the knapweed flower head. The larva emerges and burrows into the flower head where it feeds on the developing seeds. The larva damages the plant by reducing seed production (all of the seeds of diffuse knapweed and 25-100% of spotted knapweed) and the adult does damage by defoliating the plant as it feeds on the leaves prior to flowering. After flowering, adult weevils switch to feeding on flowers.

In spotted and diffuse knapweed, *L. minutus* directly consumes another biological control agent released to control the species, the gall-forming flies *Urophora affinis* Frfld and *Urophora quadrifasciata* Meigen (Diptera: Tephrididae).

This weevil is native to southern Europe and the Mediterranean. It was first released as a knapweed biocontrol in the United States in 1991 from Greece. It is now established in knapweed populations throughout the western United States, especially in warm, dry areas. It feeds on knapweeds, preferring diffuse knapweed over others, and it has not been shown to attack native flora. This weevil has been shown to reduce diffuse diffuse knapweed density at several sites. It has been less successful in controlling spotted knapweed.

Larinus minutus, a native weevil of Europe, was cleared and first released in the United States in 1991. The weevil has been released in Arizona, California, Colorado, Idaho, Minnesota, Montana (established), Nebraska, Oregon (established), South Dakota, Utah, Washington (established), and Wyoming (established) as a part of a biological control program to control spotted and diffuse knapweed (*Centaurea maculosa* Lam. and *C. diffusa* Lam.).

Appearance

L. minutus is a brown-grey weevil with a very large, bulbous snout. *L. minutus* measures 4-5 mm in length. They are strong fliers and disperse throughout the entire knapweed patch in several years. When the adults first emerge from the knapweed seedhead they are light grey with some yellow fuzz on their bodies.

Habitat

Spotted and diffuse knapweed are weed species that can be found throughout the northern tier of states and as far south as Nebraska and Virginia. These highly competitive weed species favor and establish quickly on disturbed sites and overgrazed rangeland. Both weeds will invade well established grassland communities and out compete the native vegetation. The release of *L. minutus* is part of a program to introduce a complex of spotted and diffuse knapweed enemies to help control these weeds.

Pests Attacked

L. minutus is host specific and limited to a few species in the genus *Acrolophus* and *Calcitrapa*. *L. minutus* was host tested with plants in the subgenera that are native, endangered or of economic importance before being cleared for release in the United States. *L. minutus* attacks both spotted and diffuse knapweed with a slight preference for diffuse knapweed.

Life Cycle

Adult *L. minutus* weevils emerge in late spring from the ground litter where they have been hibernating. The female weevil must feed on the flowers of the knapweed for ovary development. Therefore egg laying begins after the knapweed has started to bloom.. The eggs hatch within three days and the larvae begin feeding on the flower and migrated to the seeds and commence feeding. The seeds are consumed and when development of the larvae is complete the larva pupates and emerges for the seedhead as a new adult. Development from egg to adult takes about four weeks. The newly emerged adults feed on the plant foliage and eventually go into the soil and ground litter to hibernate for the winter.

Pesticide Susceptibility

Not yet known.

Effectiveness

L. minutus larvae destroy up to one hundred percent of the seed in an infested knapweed seedhead. This weevil along with other seed feeders will further reduce the seed that spotted and diffuse knapweed are dependent on for reproduction, dispersal, and survival.

Conservation

Sites that are chosen for *L. minutus* release should be considered for long term availability and should be left undisturbed by development or pesticide use for at least ten years. The release site should be dry with some bare ground and knapweed plants should be one to two feet apart. Frequently this will be the outer edges of the knapweed infestation.

Larinus obtusus (Coleoptera: Curculionidae)

Larinus obtusus is a species of true weevil known as the blunt knapweed flower weevil. It is used as an agent of biological pest control against noxious knapweeds, especially spotted knapweed (*Centaurea maculosa*).

The adult weevil is dark brown with a large, bulbous snout. It is 5 to 7 millimeters long in total. It is active throughout the summer when the female lays eggs in the opened flower head. The larva emerges and feeds on the developing seeds inside the head. The larval stage lasts 17 days, after which the larva constructs a cocoon from the remnants of the seeds and pupates within it for about nine days. Most of the damage to the plant is done by the larva's feeding on the seeds; the adult feeds on the foliage but does less drastic damage to the plant.

Larinus obtusus

This weevil is native to Europe and the Middle East. It was first introduced to the United States for the biocontrol of knapweeds in the Boulder, Colorado area in 1991. It is now established in much of the northwestern United States, but its abundance is unknown. This weevil is quite similar to its relative, *Larinus minutus*, which is also a knapweed specialist used for biocontrol. They are currently considered separate species but this could change with further research.

Larinus obtusus, a native of Europe, was approved and first released in 1993 in the United States. This weevil has been released in Idaho, Minnesota, Montana (established), Nebraska, Oregon, and Washington (established) as part of a program to control spotted and diffuse knapweed (*Centaurea maculosa* Lam. and *C. diffusa* Lam.).

Appearance

L. obtusus is a grey-brown weevil (5 mm) with a prominent, bulbous snout. When the adults emerge from the seedhead they are a lighter grey and have yellow fuzz on their bodies. Older *L. obtusus* tend to be darker, sometimes appearing black.

Habitat

Spotted and diffuse knapweed are weed species that can be found throughout the northern tier of states and as far south as Nebraska and Virginia. These highly competitive weed species favor and establish quickly on disturbed sites and overgrazed rangeland. Both weeds will invade well established grassland communities and out compete the native vegetation. The release of *L. obtusus* is part of a program to introduce a complex of spotted and diffuse knapweed enemies to help control these weeds.

Pests Attacked

L. obtusus was host tested against seventy-eight North American plant species including closely related, native, endangered, and economically important species. The weevil, after these tests, is considered to be host specific to *Centaurea* species in the subgenera *Acrolophus* and *Jacea*. Spotted knapweed belongs to *Acrolophus* and is one of the plants attacked by *L. obtusus*. In Europe the host plants for *L. obtusus* are *C. phrygia* (Romania) and *C. jacea* (Yugoslavia), with some attack on *C. maculosa* (spotted knapweed) in the area.

Life Cycle

L. obtusus adults emerge from the soil and litter in late spring. The adults feed on the leaves and flowers of spotted knapweed. Mating and egg laying begin with the onset of the spotted knapweed flowering. The eggs are laid in the flower, and the larvae hatch and burrow into the flower head and begin feeding on the capitula. When the larvae mature they form a hard walled "cocoon" with pappus hairs, seeds and secretions. Mating and egg laying lasts from five to six weeks after the plant starts to flower. The newly-emerged adults feed heavily on the knapweed plants and hibernate in the soil and ground litter.

Pesticide susceptibility

Not yet known

Relative Effectiveness

L. obtusus larvae destroy nearly one-hundred percent of the seeds. Any seeds that are not eaten will become a part of the "cocoon". More than one larva may develop in a seedhead.

Conservation

When choosing a site it is important to select a site that will not be disturbed by development or pesticide use for a least ten years. *L. obtusus* prefers moist sites in contrast to the other seedhead weevils for knapweed which prefer and thrive in drier sites.

Metzneria paucipunctella (Lepidoptera: Gelechiidae)

Metzneria paucipunctella is a species of moth known as the spotted knapweed seed head moth. It is used as an agent of biological pest control against noxious knapweeds, particularly spotted knapweed (*Centaurea maculosa*).

The seedhead moth, *Metzneria paucipunctella*, a native of Europe, was approved for release in 1971. *M. paucipunctella* has been released in Colorado (established), Idaho (established), Minnesota (establish-ed), Montana (established), Neb-raska, Oregon (established), and Washington (established) as a part of a complex of biocontrol agents to control spotted and diffuse knapweed (*Centaurea maculosa* and *C. diffusa*).

Metzneria paucipunctella

The adult moth is narrow-bodied and about 8 millimeters long. It is brownish-gray and lightly speckled. It has large recurved antennae. The female lays about 80 eggs, depositing each at the base of a flower head. In about ten days the larva emerges and burrows into the flower head where it feeds on the developing seeds and florets. The larva is a small, plump white grub with a dark head and visible body segments. It overwinters inside the seed head and pupates the following spring.

This moth is native to Europe. It was first introduced as a biocontrol agent for knapweeds in the United States in 1980, and it is now established in much of the western United States. The larvae of this moth compete with other head-dwelling biocontrol agents; if they find the larvae of another species they attack and consume them. Nevertheless, knapweed control is better when the moth works in tandem with other insects.

Appearance

M. paucipunctella is a small (15 mm wing span and 6 mm body length), light brown moth with a row of black spots on each front wing. This moth closely resembles many of the commonly found "grass moths". The white larva with a brown head can easily be found in the knapweed seedheads under the flower receptacle.

Habitat

Spotted and diffuse knapweed are weed species that can be found throughout the northern tier of states and as far south as Nebraska and Virginia. These highly competitive weed species favor and establish quickly on disturbed sites and overgrazed rangeland. Both weeds will invade well established grassland communities and out compete the native vegetation. The release of *M. paucipunctella* is part of a program to introduce a complex of spotted and diffuse knapweed enemies to help control these weeds.

Pests Attacked

M. paucipunctella is considered host specific to spotted knapweed (*Centaurea maculosa*). The moth has been recorded on diffuse knapweed (*Centaurea diffusa*) in Washington. Extensive host testing in Europe found the moth host specific to *C. maculosa*. Eight years of sampling of over seventy European Asteraceae found *M. paucipunctella* exclusively in *C. maculosa* seedheads. The moth larvae eat the seeds in the knapweed seedhead.

Life Cycle

M. paucipunctella produces one generation per year. The adults may be found from May, in warmer climates, to July in cooler areas. Egg laying begins two to three days after the female emerges. Each female may lay 60-100 eggs. The eggs are laid on unopened flower buds and in 10-12 days the egg hatches. The larvae then climb the flower bud, which is now opening, and enter the flower to feed inside the achenes. Later the larvae will feed on all the seeds and tunnel through the receptacle. The larvae overwinter in the seedhead and pupate in a silken chamber in the early spring. Pupation lasts about three weeks and then the adults emerge.

Pesticide Susceptibility

Not yet known.

Effectiveness

M. paucipunctella reduces seed viability and destroys most of the seeds. The moth destroys all the seed in a seedhead producing nine or fewer seeds. If the number of seeds exceeds ten the moth can be expected to destroy ninety plus percent of the seed. When the moth is present with other seedhead agents, production of seeds is markedly reduced. *M. paucipunctella* is another agent in a biocontrol agent complex to control spotted and diffuse knapweed.

Conservation

Sites chosen for *M. paucipunctella* should be available without disturbance from development or pesticide use for a least ten years. *M. paucipunctella* does not tolerate temperatures below minus 30 C. *M. paucipunctella* does well in moist as well as drier sites.

Pelochrista medullana (Lepidoptera: Tortricidae)

The Brown-winged Root Moth (*Pelochrista medullana*) is a moth of the Tortricidae family. It is native to central Europe, Turkey and southern Russia. It has been introduced to Idaho, Montana, Oregon and British Columbia.

Pelochrista medullana (Stgr.) is a root-boring moth native to central Europe that was approved for release in the United States in 1984. European studies focused on eastern Austria and eastern Romania, but it is also thought to occur in Turkey and southern Russia. Limited numbers of *P. inspersa* have been released in Idaho, Montana, Oregon and British Columbia. There is no evidence of establishment in the United States.

The wingspan is 14-21 mm. Adults are on wing from mid June to the end of July. There is one generation per year.

Larvae feed on *Centaurea maculosa, Centaurea biebersteinii* and *Centaurea diffusa* in the rosette stage. They damage the roots of their host, reducing root storage capacity and exposing the plant to pathogens. Small plants are usually completely destroyed.

Appearance

The adult is a tan to gray moth with mottled wings, measuring 8 mm in body length.

Habitat

Spotted and diffuse knapweed are invasive alien species that occur throughout the northern two-thirds of the contiguous United States. They are highly competitive weeds in semi-arid rangeland and Ponderosa Pine habitats, that establish quickly on disturbed sites and overgrazed rangeland. Both weeds can invade well-established grassland communities and displace the native vegetation. The release of *P. medullana* is part of a program to introduce a complex of natural enemies to help control these weeds.

Pests Attacked

P. medullana attacks spotted (*Centaurea maculosa* or *C. biebersteinii*) and diffuse knapweed (*C. diffusa*) in the rosette stage. Host specificity tests indicate that larvae can complete develop only on spotted and diffuse knapweed. They failed on other *Centaurea* species, including *C. jacea, C. nigrescens, C. cyanus* and *C. macrocephala*. The two native *Centaurea* spp., *C. americana* and *C. rothrockii*, are annuals which cannot support larvae, which must overwinter in the roots and complete development in the spring.

Life Cycle

P. medullana produces one generation per year. Adults emerge in July with a 1:1 sex ratio and live for about 2 weeks. Mating occurs within 24 hours of emergence, usually at dusk, and egg laying starts within 2-3 days. Eggs are laid primarily on the lower surface of rosette leaves. Females can lay up to 120 eggs in warm (18.4°C daily mean) dry weather, but this can be greatly reduced by cold (13.6°C) rainy weather. Release sites should have a mean temperature above 18°C for 3-4 weeks during the summer. Larvae hatch 7-9 days after oviposition and move to the center of the rosette and mine into the root crown. Larvae mine spiraling tunnels which are covered by a silk web in the root cortex, just under the epidermis, similar to *Agapeta zoegana*. Usually only one larva is found on an infested plant. P. medullana is known to coexist with *A. zoegana, Cleonus piger* and *Apion* spp. in Europe. Larvae overwinter in the roots and complete development in the spring.

Pesticide Susceptibility

Not known.

Effectiveness

Damage to the roots is similar to that caused by *Agapeta zoegana*. Only third- to sixth-instar larvae cause measurable damage, reducing root storage capacity and exposing the plant to pathogens. Small plants (< 5mm root diameter) are usually completely destroyed. Plants that survive insect attack are significantly smaller and produce fewer flower heads than uninfested plants.

Conservation

Sites that are chosen for *P. medullana* releases should not be disturbed for at least ten years by development or pesticide use.

Pterolonche inspersa (Lepidoptera: Pterolonchidae)

The Brown-winged knapweed Root Moth (*Pterolonche inspersa*) is a moth of the Pterolonchidae family. It is found in Portugal, Spain, France, Italy, Sardinia, the Czech Republic, Slovakia, Hungary, Bulgaria, Greece and Crete.

The root-boring moth, *Pterolonche inspersa* a native moth of Europe, was approved for release in 1986. *P. inspersa* has been released in Colorado, Montana, and Oregon. There is no known establishment in the United States of this biocontrol agent.

The wingspan is about 20 mm. Adults are light-brown without distinct markings. There is one generation per year.

The larvae feed on *Centaurea* species, including *Centaurea diffusa*. They tunnel into the root crown of their host plant and feed on the root tissue. As they reach the root cortex, they spin a silken tube and feed from within the tube. Mature larvae overwinter in the roots. In spring, a silken tube is made above the soil surface in which pupation takes place.

Appearance

The *P. inspersa* adult is a light brown moth measuring 20 mm wingspan and 8 mm in body length. There are no distinct markings on the wings.

Habitat

Spotted and diffuse knapweed are weed species that can be found throughout the northern tier of states and as far south as Nebraska and Virginia. These highly competitive weed species favor and establish quickly on disturbed sites and overgrazed rangeland. Both weeds will invade well established grassland communities and out compete the native vegetation. The release of *P. inspersa* is part of a program to introduce a complex of spotted and diffuse knapweed enemies to help control these weeds.

Pterolonche inspersa

Pests Attacked

P. inspersa prefers diffuse knapweed (*Centaurea diffusa*). The host specificity tests found the moth limited to a few *Centaurea* spp. none of which are endangered or economically important. The two native *Centaurea* spp., *C. americana* and *C. rothrockii,* are annuals and senesce and die when *P. inspersa* is laying eggs.

Life Cycle

P. inspersa produces one generation per year. Adults emerge in July and early August. The larvae overwinter in the roots of the knapweed plants. Mating and egg laying commence soon after emergence. The larvae upon hatching penetrate the root crown and begin to feed in the root. They spin a silken tube and feed from within the tube. The tube protects the larvae and provides an exit for the emerging adult. The adult lives up to fifteen days.

Pesticide Susceptibility

Not known

Effectiveness

Infested diffuse knapweed plants can be recognized by the silken tubes around the crown of the rosette. *P. inspersa* larvae cause considerable root damage and as a result the plants are stunted and produce fewer flowers. The infested root becomes spongy and the storage capacity is reduced. The roots are prone to bacterial, viral and fungal infections.

Conservation

Sites that are chosen for *P. inspersa* releases should not be disturbed for at least ten years by development and pesticide use.

Sphenoptera jugoslavica (Coleoptera: Buprestidae)

Sphenoptera jugoslavica, a root boring beetle, was first released in the United States in 1979. Its release is part of a program to introduce a complex of spotted and diffuse knapweed enemies to help control these natives of Eastern Europe and Asia. In North America spotted and diffuse knapweed are efficient competitors and since being introduced around the turn of the century have spread throughout the northern tier of states and as far south as Nebraska and Virginia. They favor disturbed sites and overgrazed rangelands and cause a reduction of forage for cattle of up to 100% in infested areas. Spotted knapweed and diffuse knapweed also prosper on grasslands and riparian sites and displace native vegetation in these habitats.

Although knapweed can be controlled by extensive cultivation or herbicide application, its existence on stream banks and rough terrain and the typically expansive areas of infestation make these methods impractical and/or economically unfeasible. Therefore, an integrated pest management approach is being sought to help in controlling this weed.

S. jugoslavica has been released in Arizona, Colorado, Idaho, Montana, Nebraska, Nevada, Oregon, South Dakota, Utah, Washington, and Wyoming. It is established in Colorado, Idaho, Montana, Oregon, South Dakota, Washington, and Wyoming, and is being collected for redistribution in Colorado, Montana, Oregon, and Washington.

Appearance

S. jugoslavica is a flat, metallic, copper colored, elongated beetle measuring 7 mm long. The larvae have an enlarged head and a long thin body. The larvae cause a gall-like swelling in the knapweed root near the crown.

Habitat

Diffuse knapweed infested areas.

Pests Attacked

European field collections found *S. jugoslavica* was limited, almost exclusively, to diffuse knapweed roots. The host-specificity tests indicated that the beetle could lay eggs and have them develop on diffuse knapweed, spotted knapweed, and yellowstar thistle.

Life Cycle

S. jugoslavica overwinters as larvae in the root of the knapweed plant. The newly formed adults remain motionless inside knapweed roots in their pupal chambers for two to five days. Then they bite a hole in the gall wall and leave the root. Adult *S. jugoslavica* emerge in July. The new adults feed on diffuse knapweed leaves for two to three days before mating. Egg laying commences in ten to twelve days after emergence. Adult *S. jugoslavica* live up to thirty days.

Pesticide Susceptibility

Not yet known.

Effectiveness

Small knapweed rosettes that have been attacked often die by the end of the next season. Diffuse knapweed plants that survive are stunted and produce less flowers. After the beetle population has built up for five to six years, the knapweed population may collapse with only scattered knapweed plants remaining. This effect has been observed in drier areas where the beetles have been released. Stunted and unhealthy plants were observed in Europe especially in drier areas.

It is anticipated that this beetle will be an effective control agent in conjunction with other root boring and seed feeding insects. Diffuse knapweed patches are hard to find in Europe because they are kept under control by the native insects and diseases.

Conservation

Long-term availability is important in sites that are chosen for *S. jugoslavia* releases. It is wise to consider sites that will be undisturbed by development or pesticide use for at least ten years. Releases should be made in sections of the diffuse knapweed infestation where the plants are at least one to two feet apart with some bare ground showing. The beetle seems to thrive in drier areas and on diffuse knapweed.

Terellia virens (Diptera: Tephritidae)

The seedhead fly, *Terellia virens*, a native of Europe, was approved for release in 1992 and has been released in California, Colorado, Idaho, Minnesota, Montana (recovered), Nebraska, Oregon (established), Washington, and Wyoming (recovered).

Appearance

T. virens is a small (3-5 mm), clear winged, greenish-brown fly. It has bright green eyes.

Habitat

Spotted knapweed (*Centaurea maculosa*) is a weed species that is found throughout the northern tier of states and as far south as Nebraska and Virginia. This highly competitive weed favors and establishes quickly on disturbed sites and overgrazed rangeland as well as established grassland communities out competing the native vegetation. The release of *T. virens* is part of a program to introduce a complex of spotted knapweed enemies to help control this weed.

Pest Attacked

T. virens has been host tested against fifty-seven plant species. The test results confirmed that *T. virens* has a limited host range and a preference for species in the *Centaurea* subgenus *Acrolophus*. Spotted knapweed (*C. maculosa*) is *T. virens* primary host. There has been very limited attack on diffuse knapweed (*C. diffusa*)

Life Cycle

The adult *T. virens* begin to emerge in late May. Weather conditions determine if there will be two generations in a season. If so, most of the second generation will overwinter in the seedheads of the mature plants as prepupal larvae. Female *T. virens* lays her eggs in young, opening flowers. Each female lays and average of 80 eggs which hatch within three to five days after oviposition. The larval development to pupation takes about fourteen days. The larvae spend their first two instars inside a single seed. Later instars feed on mature and maturing seed. A seedhead may be infested by one to seven *T. virens* larvae. The adult flies are long lived and may live for forty-eight plus days.

Pesticide Susceptibility

Not yet known.

Effectiveness

T. virens larvae cause considerable feeding damage on parts of the seed. A test for germination viability of 500 seeds was conducted. The control, no *T. virens* infestation, had 77.6% germination or 388 viable seeds compared to the *T. virens* infested seeds with 6.6% germination or 33 viable seeds. *T. virens* is released as a part of a complex of biocontrol agents for spotted knapweed will help control the weed.

Conservation

Release sites should have the potential of being free from disturbance such as development and pesticide use for at least ten years.

Urophora affinis (Diptera: Tephritidae)

Urophora affinis is a species of tephritid or fruit flies in the genus *Urophora* of the family Tephritidae. It has been released in the United States and Canada as a biocontrol agent to control spotted knapweed. *U. affinis* became established in Montana in 1973.

Urophora affinis, a seedhead fly, was first released in the United States in 1971. Its release is part of a program to introduce a complex of spotted and diffuse knapweed enemies to help control these natives of Eastern Europe and Asia. In North America spotted and diffuse knapweed are efficient competitors and since being introduced around the turn of the century have spread throughout the northern tier of states and as far south as Nebraska and Virginia. They favor disturbed sites and overgrazed rangelands and cause a reduction of forage for cattle of up to 100% in infested areas. Spotted knapweed and diffuse knapweed also prosper on grasslands and riparian sites and displace native vegetation in these habitats.

Although knapweed can be controlled by extensive cultivation or herbicide application, its existence on stream banks and rough terrain and the typically expansive areas of infestation make these methods impractical and/or economically unfeasible. Therefore, an integrated pest management approach is being sought to help in controlling this weed.

U. affinis has been released in Arizona, California, Colorado, Idaho, Michigan, Minnesota, Montana, Nebraska, Nevada, North Dakota, Oregon, South Dakota, Utah, Washington, Wisconsin, and Wyoming. Populations have been established in all of these states but Nevada. *U. affinis* may be found in the spring on the unopened buds of spotted or diffuse knapweeds.

Urophora affinis

U. affinis is multivoltine and overwinters as a larva in knapweed flower heads. In June, adults oviposit on seed heads. Each female can produce about 120 eggs. It has been shown to reduce seed production in knapweed (spotted and diffuse) by up to 95% but that is insufficient for effective management of either species. Overwintering *Urophora* larvae are heavily preyed upon by deer mice *Peromyscus maniculatus*. The inability of *Urophora* species to control knapweed led to the introduction of other biocontrol agents, including the weevils *Larinus minutus* and *Bangasternus fausti*, which also attack seed heads of kanpweed. In cage experiments, larvae of *Larinus minutus* and *Bangasternus fausti* have been shown to consume the developing *U. affinis*. Reproduction of *U. affinis* was reduced by *L. minutus*.

Appearance

U. affinis is a small (3-4 mm) fly with lightly barred wings and a dark body. The female ovipositor is prominent and nonretractable. The fly larva causes the plant to form a gall in which the larva lives, feeds, and eventually emerges as an adult. The gall on the receptacle is pointed at the tip and is made of lignified material.

Habitat

Spotted and diffuse knapweed-infested areas.

Pests Attacked

Extensive host specificity testing found that *U. affinis* is host specific to spotted and diffuse knapweed. The field and laboratory tests found that flies collected from these two weeds in various locations in Europe were host specific to the plant on which the flies were reared. Economically important plants such as artichoke and safflower have physical and visual barriers that prevent *U. affinis* from being attracted.

Life Cycle

U. affinis overwinters as larvae in the seedheads. Pupation is in the spring, and the adults emerge in mid to late spring. After mating takes place the eggs are laid between the bracts on closed flower buds. Developing larvae cause the plant to form a fusiform gall from tissues of the ovary and receptacle. The gall becomes lignified and hard. The females recognize the proper host bud by size and by probing to check the thickness and hardness of the bud exterior.

Pesticide Susceptibility

Not yet known.

Effectiveness

U. affinis creates metabolic sinks, drawing nutrients from the plant that extend beyond the attacked seedhead. New flower buds may abort and not develop due to the lack of nutrients. Unattacked seedheads produce an average of 12.6 viable seeds. Each gall is estimated to reduce seed production by 2.4 seeds per seedhead. *U. affinis* needs to be supplemented with other seedhead agents and root borers.

It is anticipated that this fly will be an effective control agent in conjunction with other root boring and seed feeding insects. Spotted and diffuse knapweed patches are hard to find in Europe because they are kept under control with the native insects and diseases.

Conservation

Release sites should be free from disturbance and from development or pesticide use for at least ten years. *U. affinis* likes full sun in open areas.

Urophora quadrifasciata (Diptera: Tephritidae)

Urophora quadrifasciata is a species of tephritid or fruit flies in the genus *Urophora* of the family Tephritidae. The host plant for the larvae is usually a knapweed (*Centaurea* sp).

Urophora quadrifasciata, a seedhead fly which is a native fly of Europe, was approved for release in 1988. *U. quadrifasciata* has been released and established in Arizona, California, Colorado, Idaho, Michigan, Minnesota, Montana, Nebraska, North Dakota, South Dakota, Utah, Washington, Wisconsin, and Wyoming. Establishment has been confirmed in Indiana, Maryland, New York, Pennsylvania, Vermont, New Hampshire, Massachusetts, Rhode Island, Connecticut, and New Jersey.

Urophora quadrifasciata

Appearance

U. quadrifasciata is a small (2-3 mm) dark bodied fly with a distinctive VU mark on its wings. The female has a long retractable ovipositor. The gall in the seedhead is paper-thin with pappus on top. The larva is white with a dark brown anal plate.

Habitat

Spotted and diffuse knapweed are weed species that can be found throughout the northern tier of states and as far south as Nebraska and Virginia. These highly competitive weed species favor and establish quickly on disturbed sites and overgrazed rangeland. Both weeds will invade well established grassland communities and out-compete the native vegetation. The release of *U. quadrifasciata* is part of a program to introduce a complex of spotted and diffuse knapweed enemies to help control these weeds.

Pests Attacked

U. quadrifasciata attacks plants in the genus *Centaurea*. The fly was tested against safflower (*Carthamus tinctorius* L.), an economically important relative of knapweed, and two US native knapweeds. No oviposition or larval development was found on these three plant species. The larvae of this fly attack the seedhead causing the plant to form a non-lignified gall around each larvae.

Life Cycle

U. quadrifasciata produce two generations per year. The overwintering generation emerges as adults from galls in the mature seedheads in the spring as the new flower buds are beginning to form. The female fly after mating, uses her ovipositor to penetrate flower buds that are over half grown to lay her eggs among the developing stamens. She may lay more than one egg per seedhead. The eggs hatch in three to four days and the larva chews down in a floret into the ovary causing the plant to start forming a gall about eight days after the larva has hatched. The gall will have reached maximum size in fifteen days after hatching. It takes about eight to nine weeks from hatching for the new adult generation to appear. The next generation or late summer generation will overwinter as mature larvae in the seedheads, emerging next spring as adults.

U. quadrifasciata is a strong flier and disperses great distances. This fly is rapidly colonizing the knapweed patches of North America.

Pesticide Susceptibility

Though high rates of 2,4-D may cause increased larval and pupal mortality, spring applications of the herbicides 2,4-D and picloram (applied when knapweeds are primarily in the rosette stage) appear to have no significant negative impacts on *U. quadrifasciata* populations.

Effectiveness

The floret that the larva occupies is destroyed and adjacent florets will abort. The gall does not form a nutrient sink and thus has a minimal impact on the total plant. Each gall displaces 1.9 seeds in a diffuse knapweed seedhead. Combined with *U. affinis,* the two flies have been reported to have destroyed up to 95% of the knapweed seed at two British Columbia sites. *U. quadrifasciata* in combination with other seedhead agents and root borers will help get diffuse and spotted knapweed under control.

Conservation

Release sites should be chosen that will not be disturbed for at least ten years by development or pesticide use. *U. quadrifasciata* prefers open sunny sites to closed shady areas.

LEAFY SPURGE CONTROL AGENTS

Aphthona abdominalis (Coleoptera: Chrysomelidae)

Aphthona abdominalis is a species of leaf beetle known as the minute spurge flea beetle. It was used as an agent of biological pest control against the noxious weed leafy spurge (*Euphorbia esula*), but never established a viable population.

The adult beetle is light orange with a black abdomen and only about 2 millimeters long. The female lays up to 100 eggs on or near leafy spurge, its host plant. The larva emerges and feeds on the roots of the plant, while the adult eats the leaves and flowers of the plant. Both larva and adult inflict damage on the plant by feeding. In its native range this beetle feeds on leafy spurge and other closely related plants. It is not known to attack other types of plants.

This beetle is native to Eurasia. It was first released as a biocontrol agent for leafy spurge in the United States in the early 1990s, but apparently never established. *A abdominalis* was never introduced to Canada. This beetle is used for biocontrol less often than other *Aphthona* species.

Leafy spurge is an Eurasian perennial that was introduced into North America in the 19th century. It infests several million hectares of rangelands and riparian areas in the United States and is a serious pest across the northern Great Plains where it displaces desirable grasses and forbs normally consumed by foraging cattle. Cattle and horses usually avoid leafy spurge, but should they eat it, its milky latex may cause sickness and even death. Annual direct and indirect economic losses due to leafy spurge infestation in Montana, South Dakota, North Dakota, and Wyoming are estimated to exceed $120 million. In addition, leafy spurge forms monocultures that often displace native plants and degrade wildlife habitats.

Leafy spurge has a number of biological characteristics that have caused it to be difficult to control with herbicides, and infestations generally occur in remote areas consisting of comparatively low-value land. Thus, classical biological control is envisioned as a potentially valuable spurge management tool in North America. To date, ten Eurasian insect species have been released as biocontrol agents of leafy spurge.

Aphthona abdominalis was originally approved for release in the United States in 1993. It has been released at several sites in Montana and North Dakota.

Appearance

The adult flea beetle is very small (1-2 mm) and yellowish-brown in color. *A. abdominalis* adults typically hop rather than fly when disturbed. Larvae are found in the soil, on or near leafy spurge roots. They are 1-3 mm long, with short legs, yellow heads, and creamy-white bodies.

Habitat

Leafy spurge-infested grasslands.

Pests attacked

The host range of *A. abdominalis* appears restricted to plants in the subgenus *Esula* of the genus *Euphorbia*. In Europe, this beetle feeds on leafy spurge and several other closely-related spurge species. There are a few native *Euphorbia* spp. in the U.S. that could potentially be hosts for *A. abdominalis,* though no feeding has yet been documented under field conditions. *A. abdominalis* will not feed on poinsettia (*Euphorbia pulcherrima*), and crop species and native plants outside the genus *Euphorbia* will not be attacked.

Life cycle

Life cycle information is based on observations in the insect's native European habitats; little information is available on the biology of *A. abdominalis* under North American field conditions.

Winter is apparently passed in the adult stage; these adults probably become active in late spring. Adults may live for several weeks to several months, feeding on leafy spurge foliage and flowers. Females lay small groups of eggs at, or just below, the soil surface, near the base of a leafy spurge stem. Newly-hatched larvae burrow into the soil and begin feeding on very small leafy spurge roots. Larvae probably feed on progressively larger roots and root buds as they develop. Pupation occurs in a soil cell near leafy spurge roots. There appears to be at least three generations per year in southern Europe, though the number of generations under North American conditions is not yet known.

Relative effectiveness

Other *Aphthona* flea beetles can kill leafy spurge plants as a direct or indirect consequence of larval feeding on spurge roots. *A. abdominalis* has the potential advantage of completing several generations per year. However, it is not yet known if this agent can survive and control leafy spurge infestations under field conditions in the U.S.

Pesticide susceptibility

Not known.

Conservation

The conservation of natural enemies is probably the most important and readily available biological control practice available to growers. Natural enemies occur in all production systems, from the backyard garden to the commercial field. They are adapted to the local environment and to the target pest, and their conservation is generally simple and cost-effective. With relatively little effort the activity of these natural enemies can be observed. Lacewings, lady beetles, hover fly larvae, and parasitized aphid mummies are almost always present in aphid colonies. Fungus-infected adult flies are often common following periods of high humidity. These natural controls are important and need to be conserved and considered when making pest management decisions. In many instances the importance of natural enemies has not been adequately studied or does not become apparent until insecticide use is stopped or reduced. Often the best we can do is to recognize that these factors are present and minimize negative impacts on them. If an insecticide is needed, every effort should be made to use a selective material in a selective manner.

Aphthona cyparissiae (Coleoptera: Chrysomelidae)

Aphthona cyparissiae is a species of leaf beetle known as the brown dot leafy spurge flea beetle. It is used as an agent of biological pest control against the noxious weed leafy spurge (*Euphorbia esula*).

The adult beetle is light golden brown and about 3 millimeters long. The female lays eggs on or near leafy spurge, its host plant, during the summer months. The white larva emerges in about two weeks and feeds on the roots and root hairs. It is active through the fall and winter until the cold stimulates it to pupate. It burrows into the soil to undergo a three-week pupation. Damage to the plant occurs when the larva eats the roots and the adult feeds on the leaves and flowers. Heavy beetle activity kills the plant. This beetle is a spurge specialist, preferring cypress spurge (*Euphorbia cyparissias*) in its native range and readily attacking leafy spurge as its first alternate.

Leafy spurge is an Eurasian perennial that was introduced into North America in the 19th century. It infests several million hectares of rangelands and riparian areas in the United States and is a serious pest across the northern Great Plains where it displaces desirable grasses and forbs normally consumed by foraging cattle. Cattle and horses usually avoid leafy spurge, but should they eat it, its milky latex may cause sickness and even death. Annual direct and indirect economic losses due to leafy spurge infestation in Montana, South Dakota, North Dakota, and Wyoming are estimated to exceed $120 million. In addition, leafy spurge forms monocultures that often displace native plants and degrade wildlife habitats.

Leafy spurge has a number of biological characteristics that have caused it to be difficult to control with herbicides, and infestations generally occur in remote areas consisting of comparatively low-value land. Thus, classical biological control is envisioned as a potentially valuable spurge management tool in North America. To date, ten Eurasian insect species have been released as biocontrol agents of leafy spurge.

Aphthona cyparissiae was originally approved for release in the United States in 1986. Through 1995, it has been widely released in leafy spurge infestations across the northern U.S. (17 states) and western Canada. Large populations are present in several states, including Wyoming, Montana, Colorado, and Wisconsin.

Appearance

Adult flea beetles are small (2-4 mm) and yellowish-brown or bronze in color above, and a darker brown below; they are very similar to the *A. nigriscutis* adults, but lack the dark spot behind the head. *A. cyparissiae* adults typically hop rather than fly when disturbed. Larvae are found in the soil, on or near leafy spurge roots. They are 1-5 mm long, with short legs, yellow heads, and creamy-white bodies.

Habitat

Leafy spurge-infested grasslands.

Pests attacked

The host range of *A. cyparissiae* appears restricted to plants in the subgenus Esula of the genus *Euphorbia*. In Europe, this beetle feeds on leafy spurge and several other closely-related spurge species. There are a few native *Euphorbia* spp. in the U.S. that could potentially be hosts for *A. cyparissiae*, though no feeding has yet been documented under field conditions. *A. cyparissiae* will not feed on poinsettia (*Euphorbia pulcherrima*), and crop species and native plants outside the genus *Euphorbia* will not be attacked.

Life Cycle

Depending on location, adults emerge from the soil beginning in June or July, and are present for several weeks to several months. *A. cyparissiae* adults feed on leafy spurge foliage and flowers, and high populations may defoliate spurge plants. Females lay small groups of eggs at, or just below, the soil surface, near the base of a leafy spurge stem. Newly-hatched larvae burrow into the soil and begin feeding on very small leafy spurge roots. Larvae feed on progressively larger roots and root buds as they develop. *A. cyparissiae* larvae overwinter, resume feeding in the spring, and then pupate in a soil cell in late spring to early summer. There is one generation per year.

Relative effectiveness

Aphthona flea beetles can kill leafy spurge plants as a direct or indirect consequence of larval feeding on spurge roots. *A. cyparissiae* populations appear to have "controlled" leafy spurge infestations at a number of western and midwestern U.S. and Canadian locations, though the overall rate of "success" is less than that for *A. nigriscutis*. This agent appears best suited to somewhat drier sites.

Pesticide susceptibility

Not known.

Conservation

The conservation of natural enemies is probably the most important and readily available biological control practice available to growers. Natural enemies occur in all production systems, from the backyard garden to the commercial field. They are adapted to the local environment and to the target pest, and their conservation is generally simple and cost-effective. With relatively little effort the activity of these natural enemies can be observed. Lacewings, lady beetles, hover fly larvae, and parasitized aphid mummies are almost always present in aphid colonies. Fungus-infected adult flies are often common following periods of high humidity. These natural controls are important and need to be conserved and considered when making pest management decisions. In many instances the importance of natural enemies has not been adequately studied or does not become apparent until insecticide use is stopped or reduced. Often the best we can do is to recognize that these factors are present and minimize negative impacts on them. If an insecticide is needed, every effort should be made to use a selective material in a selective manner.

Aphthona czwalinae (Coleoptera: Chrysomelidae)

Aphthona czwalinae is a species of leaf beetle known as the black leafy spurge flea beetle. It is used as an agent of biological pest control against the noxious weed leafy spurge (*Euphorbia esula*).

The adult beetle is shiny black and about 3 millimeters long. The female lays eggs on the soil next to leafy spurge, its host plant, during the summer months. The larva emerges in 16 or 17 days. It is white with a light brown head. It goes to work feeding on the roots of the plant throughout the winter and spring and then pupates in the soil until emerging as an adult in early summer. As the larvae weaken the roots the adults feed on the leaves and flowers of the plant, killing the plant outright or allowing infection by opportunistic fungi.

Leafy spurge is an Eurasian perennial that was introduced into North America in the 19th century. It infests several million hectares of rangelands and riparian areas in the United States and is a serious pest across the northern

This beetle is native to eastern Europe and central Asia. It was first released as a biocontrol agent for leafy spurge in the United States in the late 1980s. It was thought to be a major factor is the success of biocontrol efforts for leafy spurge until it was discovered that most of what was thought to be *A. czwalinae* was in fact *A. lacertosa*. With this realization *A. czwalinae* stopped being an important component in Aphthona leafy spurge control. *A. czwalinae* had not been verified to persist in North Dakota or Minnesota for several years, until Roehrdanz documented it in Eastern Montana in 2006

Aphthona czwalinae was originally approved for release in the United States in 1987. Through 1995, it has been widely released in mixed populations with *A. lacertosa* in leafy spurge infestations across the northern U.S. (19 states). Large populations are present in several states, including Montana, North Dakota, and Wisconsin. *Aphthona* flea beetles can kill leafy spurge plants as a direct or indirect consequence of larval feeding on spurge roots.

Appearance

Adult flea beetles are small (2-4 mm) and a shiny, metallic black in color; they are virtually indistinguishable from *A. lacertosa* adults. *A. czwalinae* adults typically hop rather than fly when disturbed. Larvae are found in the soil, on or near leafy spurge roots. They are 1-5 mm long, with short legs, yellow heads, and creamy-white bodies.

Habitat

Leafy spurge-infested grasslands.

Pests attacked

The host range of *A. czwalinae* appears restricted to plants in the subgenus *Esula* of the genus *Euphorbia*. In Europe, this beetle feeds on leafy spurge and several other closely-related spurge species. There are a few native *Euphorbia* spp. in the U.S. that could potentially be hosts for *A. czwalinae,* though no feeding has yet been documented under field conditions. *A. czwalinae* will not feed on poinsettia (*Euphorbia pulcherrima*), and crop species and native plants outside the genus *Euphorbia* will not be attacked.

Life Cycle

Depending on location, adults emerge from the soil beginning in May to July, and are present for several weeks to several months. *A. czwalinae* adults feed on leafy spurge foliage and flowers, and high populations may defoliate spurge plants. Females lay groups of eggs at, or just below, the soil surface, near the base of a leafy spurge stem. Newly-hatched larvae burrow into the soil and begin feeding on very small leafy spurge roots. Larvae feed on progressively larger roots and root buds as they develop. *A. czwalinae* larvae overwinter, resume feeding in the spring, and then pupate in a soil cell in late spring to early summer. There is one generation per year.

Relative effectiveness

Mixed *A. czwalinae*/*A. lacertosa* populations have apparently "controlled" leafy spurge infestations at a number of sites in the western and midwestern U.S. These species seem best suited to relatively mesic habitats.

Pesticide susceptibility

Not known.

Conservation

The conservation of natural enemies is probably the most important and readily available biological control practice available to growers. Natural enemies occur in all production systems, from the backyard garden to the commercial field. They are adapted to the local environment and to the target pest, and their conservation is generally simple and cost-effective. With relatively little effort the activity of these natural enemies can be observed. Lacewings, lady beetles, hover fly larvae, and parasitized aphid mummies are almost always present in aphid colonies. Fungus-infected adult flies are often common following periods of high humidity. These natural controls are important and need to be conserved and considered when making pest management decisions. In many instances the importance of natural enemies has not been adequately studied or does not become apparent until insecticide use is stopped or reduced. Often the best we can do is to recognize that these factors are present and minimize negative impacts on them. If an insecticide is needed, every effort should be made to use a selective material in a selective manner.

Aphthona flava (Coleoptera: Chrysomelidae)

Aphthona flava is a beetle of the genus *Aphthona*. It is native to Eurasia, and has been introduced into North America as a biocontrol agent for leafy spurge. Adults are brown and difficult to distinguish in the field from *A. cyparissiae* and *A. nigriscutis*. *A. flava* is still the dominant beetle at some Manitoba release sites, but it is relatively scarce in North Dakota and Minnesota. Besides Manitoba, it is widespread in Hungary and Italy.

Aphthona flava

Leafy spurge is an Eurasian perennial that was introduced into North America in the 19th century.

It infests several million hectares of rangelands and riparian areas in the United States and is a serious pest across the northern Great Plains where it displaces desirable grasses and forbs normally consumed by foraging cattle. Cattle and horses usually avoid leafy spurge, but should they eat it, its milky latex may cause sickness and even death. Annual direct and indirect economic losses due to leafy spurge infestation in Montana, South Dakota, North Dakota, and Wyoming are estimated to exceed $120 million. In addition, leafy spurge forms monocultures that often displace native plants and degrade wildlife habitats.

Leafy spurge has a number of biological characteristics that have caused it to be difficult to control with herbicides, and infestations generally occur in remote areas consisting of comparatively low-value land. Thus, classical biological control is envisioned as a potentially valuable spurge management tool in North America. To date, ten Eurasian insect species have been released as biocontrol agents of leafy spurge.

Aphthona flava was originally approved for release in the United States in1986. Through 1995, it has been widely released in leafy spurge infestations across the northern U.S. (19 states) and western Canada. Large populations are present in several states, including Montana and North Dakota. *Aphthona* flea beetles can kill leafy spurge plants as a direct or indirect consequence of larval feeding on spurge roots.

Appearance

Adult flea beetles are small (3-4 mm) and orange or orange-brown in color. *A. flava* adults typically hop rather than fly when disturbed. Larvae are found in the soil, on or near leafy spurge roots. They are 1-6 mm long, with short legs, yellow heads, and creamy-white bodies.

Habitat

Leafy spurge-infested grasslands.

Pests attacked

The host range of *A. flava* appears restricted to plants in the subgenus *Esula* of the genus *Euphorbia*. In Europe, this beetle feeds on leafy spurge and several other closely-related spurge species. There are a few native *Euphorbia* spp. in the U.S. that could potentially be hosts for *A. flava*, though no feeding has yet been documented under field conditions. *A. flava* will not feed on poinsettia (*Euphorbia pulcherrima*), and crop species and native plants outside the genus *Euphorbia* will not be attacked.

Life Cycle

Depending on location, adults emerge from the soil beginning in June to August, and are present for several weeks to several months. *A. flava* adults feed on leafy spurge foliage and flowers, and high populations may defoliate spurge plants. Females lay groups of 20-30 eggs at, or just below, the soil surface, near the base of a leafy spurge stem. Newly-hatched larvae burrow into the soil and begin feeding on very small leafy spurge roots. Larvae feed on progressively larger roots and root buds as they develop. *A. flava* larvae overwinter, resume feeding in the spring, and then pupate in a soil cell in late spring to early summer. There is one generation per year.

Larvae

The larvae is active from July to early spring of next year. When they are young, they feed on the hairs of the roots. As soon as they grow up they become bigger and start eating bigger roots. The larvae size is tiny, and are of white colour and are worm-like.

Pupa

Starting from late spring to early summer the pupa stage begins in the soil.

Adult

From June to through early fall, if the temperature is right, the adults will come out. They look like like flea beetles, and behave like ones too. Their size is 3.4mm for males, and 3.6mm for females.

Relative effectiveness

A. flava populations appear to have succesfully managed leafy spurge infestations at several locations, but the overall "success rate" is lower than that for other flea beetle species. This species seems best suited to sites that are somewhat more mesic that those utilized by *A. nigriscutis* or *A. cyparissiae*.

Pesticide susceptibility

Not known.

Conservation

The conservation of natural enemies is probably the most important and readily available biological control practice available to growers. Natural enemies occur in all production systems, from the backyard garden to the commercial field. They are adapted to the local environment and to the target pest, and their conservation is generally simple and cost-effective. With relatively little effort the activity of these natural enemies can be observed. Lacewings, lady beetles, hover fly larvae, and parasitized aphid mummies are almost always present in aphid colonies. Fungus-infected adult flies are often common following periods of high humidity. These natural controls are important and need to be conserved and considered when making pest management decisions. In many instances the importance of natural enemies has not been adequately studied or does not become apparent until insecticide use is stopped or reduced. Often the best we can do is to recognize that these factors are present and minimize negative impacts on them. If an insecticide is needed, every effort should be made to use a selective material in a selective manner.

Aphthona lacertosa (Coleoptera: Chrysomelidae)

Aphthona lacertosa is a root-feeding flea beetle of the genus *Aphthona*. It is one of 5 *Aphthona* spp. that has been used in Alberta, Canada to control leafy spurge, an invasive plant that reduces pasture quality and degrades natural habitats.

A. lacertosa is native to Eurasia, but was released into North America first in 1990, then more widely in 1997. *A. lacertosa* was released specifically to control leafy spurge in wet habitats, where the use of herbicides is restricted and other previous biocontrol efforts had proven ineffective. It has established in Manitoba, Saskatchewan, and Alberta.

Adults of this species are black and indistinguishable from *A. czwalinae* in the field.

Larvae cause the most damage to the plant by feeding on the roots. In one study, beetle success in controlling leafy spurge was related to the temperature degree days at the control site, with sites with more temperature degree days having more success

Leafy spurge is an Eurasian perennial that was introduced into North America in the 19th century. It infests several million hectares of rangelands and riparian areas in the United States and is a serious pest across the northern Great Plains where it displaces desirable grasses and forbs normally consumed by foraging cattle. Cattle and horses usually avoid leafy spurge, but should they eat it, its milky latex may cause sickness and even death. Annual direct and indirect economic losses due to leafy spurge infestation in Montana, South Dakota, North Dakota, and Wyoming are estimated to exceed $120 million. In addition, leafy spurge forms monocultures that often displace native plants and degrade wildlife habitats.

Leafy spurge has a number of biological characteristics that have caused it to be difficult to control with herbicides, and infestations generally occur in remote areas consisting of comparatively low-value land. Thus, classical biological control is envisioned as a potentially valuable spurge management tool in North America. To date, ten Eurasian insect species have been released as biocontrol agents of leafy spurge.

Appearance

Adult flea beetles are small (2-4 mm) and a shiny, metallic black in color; they are virtually indistinguishable from *A. czwalinae* adults. *A. lacertosa* adults typically hop rather than fly when disturbed. Larvae are found in the soil, on or near leafy spurge roots. They are 1-5 mm long, with short legs, yellow heads, and creamy-white bodies.

Habitat

Leafy spurge-infested grasslands.

Pests attacked

The host range of *A. lacertosa* appears restricted to plants in the subgenus *Esula* of the genus *Euphorbia*. In Europe, this beetle feeds on leafy spurge and several other closely-related spurge species. There are a few native *Euphorbia* spp. in the U.S. that could potentially be hosts for *A. lacertosa,* though no feeding has yet been documented under field conditions. *A. lacertosa* will not feed on poinsettia (*Euphorbia pulcherrima*), and crop species and native plants outside the genus *Euphorbia* will not be attacked.

Life Cycle

Depending on location, adults emerge from the soil beginning in May to July, and are present for several weeks to several months. *A. lacertosa* adults feed on leafy spurge foliage and flowers, and high populations may defoliate spurge plants. Females lay groups of eggs at, or just below, the soil surface, near the base of a leafy spurge stem. Newly-hatched larvae burrow into the soil and begin feeding on very small leafy spurge roots. Larvae feed on progressively larger roots and root buds as they develop. *A. lacertosa* larvae overwinter, resume feeding in the spring, and then pupate in a soil cell in late spring to early summer. There is one generation per year.

Relative effectiveness

Mixed A. czwalinae/*A. lacertosa* populations have apparently "controlled" leafy spurge infestations at a number of sites in the western and midwestern U.S. These species seem best suited to relatively mesic habitats.

Pesticide susceptibility

Not known.

Conservation

The conservation of natural enemies is probably the most important and readily available biological control practice available to growers. Natural enemies occur in all production systems, from the backyard garden to the commercial field. They are adapted to the local environment and to the target pest, and their conservation is generally simple and cost-effective. With relatively little effort the activity of these natural enemies can be observed. Lacewings, lady beetles, hover fly larvae, and parasitized aphid mummies are almost always present in aphid colonies. Fungus-infected adult flies are often common following periods of high humidity. These natural controls are important and need to be conserved and considered when making pest management decisions. In many instances the importance of natural enemies has not been adequately studied or does not become apparent until insecticide use is stopped or reduced. Often the best we can do is to recognize that these factors are present and minimize negative impacts on them. If an insecticide is needed, every effort should be made to use a selective material in a selective manner.

Aphthona nigriscutis (Coleoptera: Chrysomelidae)

Aphthona nigriscutis is a root-feeding flea beetle of the genus *Aphthona*. It is one of 5 *Aphthona* spp. that has been used in Alberta, Canada to control leafy spurge, an invasive plant that reduces pasture quality and degrades natural habitats. It, along with *A. lacertosa,* is one of only two biocontrol agents thought to be effective against leafy spurge.

Larvae cause the most damage to the plant by feeding on the roots.

Leafy spurge is an Eurasian perennial that was introduced into North America in the 19th century. It infests several million hectares of rangelands and riparian areas in the United States and is a serious pest across the northern Great Plains where it displaces desirable grasses and forbs normally consumed by foraging cattle. Cattle and horses usually avoid leafy spurge, but should they eat it, its milky latex may cause sickness and even death. Annual direct and indirect economic losses due to leafy spurge infestation in Montana, South Dakota, North Dakota, and Wyoming are estimated to exceed $120 million. In addition, leafy spurge forms monocultures that often displace native plants and degrade wildlife habitats.

Leafy spurge has a number of biological characteristics that have caused it to be difficult to control with herbicides, and infestations generally occur in remote areas consisting of comparatively low-value land. Thus, classical biological control is envisioned as a potentially valuable spurge management tool in North America. To date, ten Eurasian insect species have been released as biocontrol agents of leafy spurge.

Aphthona nigriscutis was originally approved for release in the United States in 1989. Through 1995, it has been widely released in leafy spurge infestations across the northern U.S. (19 states) and western Canada. Large populations are present in several western states, including Wyoming, Montana, North Dakota, and Colorado. *Aphthona* flea beetles can kill leafy spurge plants as a direct or indirect consequence of larval feeding on spurge roots.

Appearance

The adult flea beetle is small (2-4 mm) and yellowish-brown or bronze in color above, and a darker brown below. A black spot is present on the dorsal surface, behind the head. *A. nigriscutis* adults typically hop rather than fly when disturbed. Larvae are found in the soil, on or near leafy spurge roots. They are 1-5 mm long, with short legs, yellow heads, and creamy-white bodies.

Habitat

Leafy spurge-infested grasslands.

Pests attacked

The host range of *A. nigriscutis* appears restricted to plants in the subgenus *Esula* of the genus *Euphorbia*. In Europe, this beetle feeds on leafy spurge and several other closely-related spurge species. There are a few native *Euphorbia* spp. in the U.S. that could potentially be hosts for *A. nigriscutis*, though no feeding has yet been documented under field conditions. *A. nigriscutis* will not feed on poinsettia (*Euphorbia pulcherrima*), and crop species and native plants outside the genus Euphorbia will not be attacked.

Life cycle

Depending on location, adults emerge from the soil beginning in May to July, and are present for several weeks to several months. *A. nigriscutis* adults feed on leafy spurge foliage and flowers, and high populations may defoliate spurge plants. Females lay small groups of eggs at, or just below, the soil surface, near the base of a leafy spurge stem. Newly-hatched larvae burrow into the soil and begin feeding on very small leafy spurge roots. Larvae feed on progressively larger roots and root buds as they develop. *A. nigriscutis* larvae overwinter, resume feeding in the spring, and then pupate in a soil cell in late spring to early summer. There is one generation per year.

Relative effectiveness

A. nigriscutis populations have significantly reduced spurge infestations over large areas in several western states and in Canada. Drier sites appear optimal for this species.

Pesticide susceptibility

Not yet known.

Conservation

The conservation of natural enemies is probably the most important and readily available biological control practice available to growers. Natural enemies occur in all production systems, from the backyard garden to the commercial field. They are adapted to the local environment and to the target pest, and their conservation is generally simple and cost-effective. With relatively little effort the activity of these natural enemies can be observed. Lacewings, lady beetles, hover fly larvae, and parasitized aphid mummies are almost always present in aphid colonies. Fungus-infected adult flies are often common following periods of high humidity. These natural controls are important and need to be conserved and considered when making pest management decisions. In many instances the importance of natural enemies has not been adequately studied or does not become apparent until insecticide use is stopped or reduced. Often the best we can do is to recognize that these factors are present and minimize negative impacts on them. If an insecticide is needed, every effort should be made to use a selective material in a selective manner.

Chamaesphecia hungarica (Lepidoptera: Sesiidae)

The Hungarian Clearwing Moth (*Chamaesphecia hungarica*) is a moth of the Sesiidae family. It is native to the south-eastern Czech Republic and Slovakia, Austria, Hungary, Serbia and Croatia. It was originally approved for introduction into the United States in 1993. It has been released at several leafy spurge-infested sites in Montana and North Dakota.

Chamaesphecia hungarica

Adults are 10-14 mm in length. They are dark brown with yellow markings. The wings are mostly clear, with dark margins and brown and yellow markings. In Europe it is usually found in moist loamy soils and partial shade near riverbanks, swampy areas and ditches. Here, adults emerge from mid May to late July.

The larvae feed on plants in the subgenus *Esula* of the genus *Euphorbia*. The larvae mine the stem and move into the pith area. They overwinter in the roots and migrate to the stem base in spring where pupation occurs.

Leafy spurge is an Eurasian perennial that was introduced into North America in the 19th century. It infests several million hectares of rangelands and riparian areas in the United States and is a serious pest across the northern Great Plains where it displaces desirable grasses and forbs normally consumed by foraging cattle. Cattle and horses usually avoid leafy spurge, but should they eat it, its milky latex may cause sickness and even death. Annual direct and indirect economic losses due to leafy spurge infestation in Montana, South Dakota, North Dakota, and Wyoming are estimated to exceed $120 million. In addition, leafy spurge forms monocultures that often displace native plants and degrade wildlife habitats.

Leafy spurge has a number of biological characteristics that have caused it to be difficult to control with herbicides, and infestations generally occur in remote areas consisting of comparatively low-value land. Thus, classical biological control is envisioned as a potentially valuable spurge management tool in North America. To date, ten Eurasian insect species have been released as biocontrol agents of leafy spurge.

Chamaespecia hungarica was originally approved for introduction into the United States in 1993. It has been released at several leafy spurge-infested sites in Montana and North Dakota.

Appearance

The adult moth is 10-14 mm in length, and generally dark brown with yellow markings. The wings are mostly clear, with dark margins and brown and yellow markings. Antennae are about half as long as the body and dark; the legs are yellow with brown markings. *C. hungarica* larvae are found in leafy spurge roots. They are typical caterpillars reaching up to 15 mm in length; the body is generally white or somewhat yellowish in color.

Habitat

Leafy spurge-infested grasslands.

Pests attacked

The host range of *C. hungarica* appears restricted to plants in the subgenus *Esula* of the genus *Euphorbia*. In Europe, this beetle feeds on leafy spurge and several other closely-related spurge species. There are a few native *Euphorbia* spp. in the U.S. that could potentially be hosts for *C. hungarica*, though no feeding has yet been documented under field conditions. *C. hungarica* will not feed on poinsettia (*Euphorbia pulcherrima*), and crop species and native plants outside the genus Euphorbia will not be attacked.

Life cycle

Life cycle information is based on observations in the insect's native European habitats; little information on the biology of *C. hungarica* under North American field conditions is available.

Adult moths are present in early to mid-summer. After mating, females lay single eggs on leafy spurge stems, leaves, and flowers; generally, 100-200 eggs are laid by each female. Newly-hatched larvae bore into a spurge stem and tunnel down the stem into the roots. Larvae feed within the roots, overwinter, and resume development the following spring. In late spring, the mature caterpillar tunnels up inside a leafy spurge stem until slightly above the soil surface, chews an exit hole, and pupates within the stem. The newly-emerged adult moth then crawls out through the exit hole. There is one generation per year in central Europe.

Relative effectiveness

In time, this species may be a useful biological control agent because it is a root feeder and because, unlike other available biocontrol agents, it may be adapted to moist and shaded habitats. However, information on its survival and host impacts under U.S. field conditions is not yet available.

Pesticide susceptibility

Not known

Conservation

The conservation of natural enemies is probably the most important and readily available biological control practice available to growers. Natural enemies occur in all production systems, from the backyard garden to the commercial field. They are adapted to the local environment and to the target pest, and their conservation is generally simple and cost-effective. With relatively little effort the activity of these natural enemies can be observed. Lacewings, lady beetles, hover fly larvae, and parasitized aphid mummies are almost always present in aphid colonies. Fungus-infected adult flies are often common following periods of high humidity. These natural controls are important and need to be conserved and considered when making pest management decisions. In many instances the importance of natural enemies has not been adequately studied or does not become apparent until insecticide use is stopped or reduced. Often the best we can do is to recognize that these factors are present and minimize negative impacts on them. If an insecticide is needed, every effort should be made to use a selective material in a selective manner.

Hyles euphorbiae (Lepidoptera: Sphingidae)

The Spurge Hawk-moth (*Hyles euphorbiae*) is a European moth of the family Sphingidae. This hawk moth is used as an agent of biological pest control against the noxious weed known as leafy spurge (*Euphorbia esula*), but usually only in conjunction with other agents. The larvae consume the leaves and bracts of the plant.

Hyles euphorbiae

Leafy spurge is an Eurasian perennial that was introduced into North America in the 19th century.

It infests several million hectares of rangelands and riparian areas in the United States and is a serious pest across the northern Great Plains where it displaces desirable grasses and forbs normally consumed by foraging cattle. Cattle and horses usually avoid leafy spurge, but should they eat it, its milky latex may cause sickness and even death. Annual direct and indirect economic losses due to leafy spurge infestation in Montana, South Dakota, North Dakota, and Wyoming are estimated to exceed $120 million. In addition, leafy spurge forms monocultures that often displace native plants and degrade wildlife habitats.

Leafy spurge has a number of biological characteristics that have caused it to be difficult to control with herbicides, and infestations generally occur in remote areas consisting of comparatively low-value land. Thus, classical biological control is envisioned as a potentially valuable spurge management tool in North America. To date, ten Eurasian insect species have been released as biocontrol agents of leafy spurge.

The leafy spurge hawk moth was the first classical biological agent released against leafy spurge in the United States, with approval for introduction granted in 1965. Populations of this insect are present in several western states, including Montana, Idaho, and Oregon.

Description

"The Spurge Hawk-Moth - Fore wings grey, with an almost square olive-brown blotch; at the base another olive-brown blotch near the middle, and a long oblique band of the same colour, commencing in a point at the extreme apex of the wing, and gradually growing wider until it reaches the margin, where it is very broad: hind wings pink, with a black blotch at the base, and a black band half-way between this black blotch and the margin, and a snowy-white blotch at the anal angle: thorax and body olive-brown, with a white line on each side of the thorax just at the base of the wings; this line runs on each side along the head just above the eye, and the two meet at the nose; the body has on each side at the base two square black spots and two square white spots, and beyond them, nearer the apex, and also on each side, are three white lines.

The caterpillar is smooth and black, with innumerable whitish dots; there are also eleven large spots of the same colour arrayed in a row on each side of the back, and beneath these as many spots of the same size and of a bright coral-red colour; the head is of the same coral-red colour, and a line of the same colour runs all along the back, from the head to the horn; the horn is red at the base and black at the tip. It feeds on sea-spurge. The chrysalis is pale brown and delicately lined and dotted with black in the manner of network; it buries itself in the loose dry sand on the sea coast.

The eggs are covered with liquid gum, which enables them to stick on the small leaves of the spurge. In a fortnight these hatch and produce little black caterpillars; the white and red spots appear as the caterpillar increases in size, and in a few weeks it becomes a most beautiful object, and so conspicuous as to attract the sea-gulls and terns, which devour them in numbers. We have never had the pleasure of finding either the caterpillar or perfect moth.

Appearance

Adult moths are large (length: 2-3 cm, wingspan: 5-7 cm) day- flying moths that often exhibit a hummingbird-like flight while visiting flowers. The body is light brown with various white and dark brown markings, while the wings have a conspicuous tan, brown, and pink or red color pattern.

H. euphorbiae larvae are found feeding on leafy spurge leaves. The caterpillars are also conspicuously colored, with a pronounced tail or "horn" near the rear end. Young larvae are variously patterned with green, yellow, and black; older larvae have a distinctive red, black, yellow, and white color pattern. Mature larvae may approach 10 cm in length; when disturbed, they regurgitate a slimy green liquid. Pupae are 3.5-5 cm long and dark brown, and are found in the soil.

Habitat

Leafy spurge-infested grasslands.

Pests attacked

The host range of *H. euphorbiae* appears restricted to plants in the subgenus *Esula* of the genus *Euphorbia*. In Europe, this caterpillar feeds on leafy spurge and several other closely-related spurge species. There are a few native *Euphorbia* spp. in the U.S. that could potentially be hosts for *H. euphorbiae*, though no feeding has yet been documented under field conditions. *H. euphorbiae* will not feed on poinsettia (*Euphorbia pulcherrima*), and crop species and native plants outside the genus *Euphorbia* will not be attacked.

Life Cycle

Adult moths are present beginning in early to mid-summer. After mating, females lay small clusters of eggs on leafy spurge foliage. After hatching, larvae consume leafy spurge leaves and flowers. Mature larvae enter the soil to pupate. There are one or two generations per year, with soil-inhabiting pupae as the overwintering stage.

Relative effectiveness

H. euphorbiae larvae defoliate leafy spurge plants, but this damage seems to have little or no impact on spurge populations. In addition, hawk moth populations generally remain low in an area, due to predation and disease. Thus, *H. euphorbiae* plays only a very minor role in leafy spurge biological control.

Pesticide susceptibility

The herbicide 2,4-D appears to have no direct impact on *H. euphorbiae* populations.

Conservation

The conservation of natural enemies is probably the most important and readily available biological control practice available to growers. Natural enemies occur in all production systems, from the backyard garden to the commercial field. They are adapted to the local environment and to the target pest, and their conservation is generally simple and cost-effective. With relatively little effort the activity of these natural enemies can be observed. Lacewings, lady beetles, hover fly larvae, and parasitized aphid mummies are almost always present in aphid colonies. Fungus-infected adult flies are often common following periods of high humidity. These natural controls are important and need to be conserved and considered when making pest management decisions. In many instances the importance of natural enemies has not been adequately studied or does not become apparent until insecticide use is stopped or reduced. Often the best we can do is to recognize that these factors are present and minimize negative impacts on them. If an insecticide is needed, every effort should be made to use a selective material in a selective manner.

Oberea erythrocephala (Coleoptera: Cerambycidae)

Leafy spurge is an Eurasian perennial that was introduced into North America in the 19th century. It infests several million hectares of rangelands and riparian

areas in the United States and is a serious pest across the northern Great Plains where it displaces desirable grasses and forbs normally consumed by foraging cattle. Cattle and horses usually avoid leafy spurge, but should they eat it, its milky latex may cause sickness and even death. Annual direct and indirect economic losses due to leafy spurge infestation in Montana, South Dakota, North Dakota, and Wyoming are estimated to exceed $120 million. In addition, leafy spurge forms monocultures that often displace native plants and degrade wildlife habitats.

Leafy spurge has a number of biological characteristics that have caused it to be difficult to control with herbicides, and infestations generally occur in remote areas consisting of comparatively low-value land. Thus, classical biological control is envisioned as a potentially valuable spurge management tool in North America. To date, ten Eurasian insect species have been released as biocontrol agents of leafy spurge.

Oberea erythrocephala was initially approved for introduction into the United States in 1980. Through 1995, it has been released at leafy spurge-infested sites in at least 15 states across the northern U.S. and in western Canada.

Appearance

Oberea erythrocephala adults are slender beetles 10-12 mm in length, with long, dark antennae. The beetles are slate-gray above with a reddish-orange head, and light grey with reddish markings below; the legs are yellowish-brown. They are active fliers, and may be seen moving above the leafy spurge canopy. The legless larvae are found within leafy spurge roots, and may reach a length of 20 mm. They are white with a yellowish head, and the body is obviously segmented.

Habitat

Leafy spurge-infested grasslands.

Pests attacked

The host range of *O. erythrocephala* appears restricted to plants in the subgenus *Esula* of the genus *Euphorbia*. In Europe, this beetle feeds on leafy spurge and several other closely-related spurge species. There are a few native *Euphorbia* spp. in the U.S. that could potentially be hosts for *O. erythrocephala,* though no feeding has yet been documented under field conditions. *O. erythrocephala* will not feed on poinsettia (*Euphorbia pulcherrima*), and crop species and native plants outside the genus *Euphorbia* will not be attacked.

Life Cycle

Adult beetles are present in early to mid-summer, and feed on leafy spurge leaves and flowers. After mating, the female beetle chews a hole in the upper part of a leafy spurge stem, into which an egg is deposited. Generally, a single egg is laid per stem, but a female may lay a total of 40 eggs. After hatching, young larvae tunnel down the leafy spurge stem until reaching the root crown area, just below the soil surface. Larvae do most of their feeding within the root crown and larger lateral roots.

Winter is spent in the larval stage, and development resumes in the spring. Larvae construct a cell within the upper root crown in which pupation occurs. Newly-eclosed adults chew through remaining root tissue and emerge from the soil. There is one generation per year in Europe, but it appears that two years may be required to complete the life cycle in the northern U.S.

Relative effectiveness

This species is potentially a valuable biological control agent because of the extensive damage it can cause in leafy spurge root systems. However, there is not yet sufficient information available from North American populations to determine the efficacy of this agent.

Pesticide susceptibility

Not known.

Conservation

The conservation of natural enemies is probably the most important and readily available biological control practice available to growers. Natural enemies occur in all production systems, from the backyard garden to the commercial field. They are adapted to the local environment and to the target pest, and their conservation is generally simple and cost-effective. With relatively little effort the activity of these natural enemies can be observed. Lacewings, lady beetles, hover fly larvae, and parasitized aphid mummies are almost always present in aphid colonies. Fungus-infected adult flies are often common following periods of high humidity. These natural controls are important and need to be conserved and considered when making pest management decisions. In many instances the importance of natural enemies has not been adequately studied or does not become apparent until insecticide use is stopped or reduced. Often the best we can do is to recognize that these factors are present and minimize negative impacts on them. If an insecticide is needed, every effort should be made to use a selective material in a selective manner.

Spurgia esulae (Diptera: Cecidomyiidae)

Leafy spurge is an Eurasian perennial that was introduced into North America in the 19th century. It infests several million hectares of rangelands and riparian areas in the United States and is a serious pest across the northern Great Plains where it displaces desirable grasses and forbs normally consumed by foraging cattle. Cattle and horses usually avoid leafy spurge, but should they eat it, its milky latex may cause sickness and even death. Annual direct and indirect economic losses due to leafy spurge infestation in Montana, South Dakota, North Dakota, and Wyoming are estimated to exceed $120 million. In addition, leafy spurge forms monocultures that often displace native plants and degrade wildlife habitats.

Leafy spurge has a number of biological characteristics that have caused it to be difficult to control with herbicides, and infestations generally occur in remote areas consisting of comparatively low-value land. Thus, classical biological control is envisioned as a potentially valuable spurge management tool in North America. To date, ten Eurasian insect species have been released as biocontrol agents of leafy spurge.

Spurgia esulae was first approved for introduction into the United States in 1985. Through 1995, *S. esulae* has been released at leafy spurge-infested sites in 19 states across the northern U.S.

Appearance

Spurgia esulae adults are very small (1-2 mm) dark-grey flies with a reddish abdomen, that are somewhat mosquito-like in appearance. Galls produced by *S. esulae* on leafy spurge stems resemble small cabbages or artichokes, with overlapping warty, pale-green leaves. They are quite variable in size, generally ranging from 6-40 mm in length. The legless orange larvae (1-2 mm) are found within the galls.

Habitat

Leafy spurge-infested grasslands.

Pests attacked

The host range of *S. esulae* appears restricted to plants in the subgenus *Esula* of the genus *Euphorbia*. In Europe, this beetle feeds on leafy spurge and several other closely-related spurge species. There are a few native

Euphorbia spp. in the U.S. that could potentially be hosts for *S. esulae*, though no feeding has yet been documented under field conditions. *S. esulae* will not feed on poinsettia (*Euphorbia pulcherrima*), and crop species and native plants outside the genus *Euphorbia* will not be attacked.

Life Cycle

Spurgia esulae adults first appear in late spring to early summer, but are often difficult to find due to their small size and short life span. After mating, females lay groups of 20 or more small (<1 mm) orange eggs on leafy spurge stems or leaves near an apical bud. Newly-hatched larvae crawl into the bud tissues and begin feeding, which causes the plant to produce the gall. Larvae feed and develop within the galls; mature larvae construct silken cocoons within which pupation occurs.

S. esulae appears to complete from two to five generations per year depending on location. During earlier generations, pupation occurs within, and adults emerge from, the mature galls. During the last generation, however, larvae exit the galls, enter the soil, and overwinter. Pupation occurs the following spring, and adults emerge from the soil.

Relative effectiveness

Galls produced by the plant in response to *S. esulae* feeding rarely kill a leafy spurge plant. However, flower and, hence, seed production may be reduced at high gall densities.

Pesticide susceptibility

The herbicides 2,4-D, picloram, and imazethapyr appear to have no impact on *S. esulae* populations.

Conservation

The conservation of natural enemies is probably the most important and readily available biological control practice available to growers. Natural enemies occur in all production systems, from the backyard garden to the commercial field. They are adapted to the local environment and to the target pest, and their conservation is generally simple and cost-effective. With relatively little effort the activity of these natural enemies can be observed. Lacewings, lady beetles, hover fly larvae, and parasitized aphid mummies are almost always present in aphid colonies. Fungus-infected adult flies are often common following periods of high humidity. These natural controls are important and need to be conserved and considered when making pest management decisions. In many instances the importance of natural enemies has not been adequately studied or does not become apparent until insecticide use is stopped or reduced. Often the best we can do is to recognize that these factors are present and minimize negative impacts on them. If an insecticide is needed, every effort should be made to use a selective material in a selective manner.

MELALEUCA CONTROL AGENT

Oxyops vitiosa Pascoe (Coleoptera: Curculionidae)

Oxyops vitiosa is a species of weevil in the Curculionidae family. Common names include the melaleuca leaf weevil and the melaleuca snout beetle. It feeds on the leaves and shoots of the broad-leaved paper bark tree, *Melaleuca quinquenervia,* which is endemic to Australia where it grows on seasonally inundated plains and swampland. It was introduced into Florida in order to help drain flooded portions of the Everglades.

Oxyops vitiosa and its host, *Melaleuca quinquenervia,* which is an invasive weedy tree, are native to Australia. The tree, known as melaleuca or the paperbark tree, was imported in 1906 and later planted extensively to drain the Everglades. Melaleuca is now considered to be destroying the Everglades, spreading to 50 additional acres of wetlands every day.

Appearance

Adult melaleuca weevils are small, 6-9 mm long, and gray to brown. Eggs are covered with a secretion that dries chocolate to black in color and gives them a teardrop shape. Without this secretion, eggs are yellow, 1 mm long, and resemble a gelatin capsule. However, they are rarely found uncovered. Larvae range in size from less than 1 mm when newly emerged to 14 mm in the fourth instar. Newly hatched larvae are yellow but appear brown to black once feeding begins. They are usually covered with a sticky secretion or exudate, and, especially on older larvae, a long, thin coil of feces is prominent. The pupal stage takes place underground in pupal capsules formed from the soil.

Adults are inconspicuous on the leaves and stems of young seedlings and the young growth of saplings and older trees, and their presence can be determined most easily by scouting for adult feeding damage. This appears as holes 1-2 mm in diameter to elongate holes, 2-3 mm wide x 5-20 mm long. Occasionally, when feeding is superficial on leaves, trenches rather than holes are evident. On stems, feeding damage consists of trenches, short or long, and usually along one side. Sometimes when stem feeding occurs on very young growth, the stem is nearly excised causing the upper portion to fall over and hang down. Adults may feed superficially on buds, leaving holes in the outer portion, or may feed deeply, leaving pits.

When eggs are present, they are likely to be near or next to the feeding damage. Eggs usually will be layed singly or in small groups on the tips of young leaves and are sometimes also found on the upper and lower surfaces of leaves, leaf buds, and on tender stems.

Larval feeding damage differs from that of adults. Larvae eat through all layers of the leaf except the cuticle on the opposite side and produce paper-thin trenches in the leaves about the width of their bodies. This larval damage provides the most characteristic evidence of melaleuca weevil infestation.

Habitat

Stands of *Melaleuca quinquenervia,* the paperbark tree, which is well established in central and south Florida.

Pests Attacked

Melaleuca quinquenervia, the paperbark tree which is in the family Myrtaceae. Testing in quarantine by G. Buckingham of over 100 plant species with adult *O. vitiosa* and of 30 species (including all native Florida Myrtaceae) with larvae and/or ovipositing females demonstrated that the weevil will establish populations only on melaleuca. Wandering adults and larvae might occasionally feed on young leaves of the native wax myrtle, *Myrica cerifera* and of some Myrtaceae (mostly imported bottlebrushes and eucalyptus), but feeding is transitory.

Researcher conducted extensive field surveys in Australia prior to the Florida testing in quarantine. No more than 7 individuals, including all stages, were found in the field on non-host species (bottlebrush, eucalyptus, and other melaleucas) compared with more than 1000 on melaleuca. In Balciunas' laboratory tests, adults were produced only on melaleuca.

The family Myrtaceae includes 8 species native to Florida in the genera *Eugenia, Psidium, Myrcianthes,* and *Calyptranthes*. Many imported species are cultivated in Florida for the fruit (*Eugenia, Myciaria, Psidium,* and *Syzygium*) and for landscaping (*Eucalyptus* and the bottlebrushes, *Callistemon*).

Life Cycle

Adults feed, mate, and lay eggs on the leaves and stems of young seedlings and the new growth of older plants. Newly hatched larvae begin feeding just outside the eggs. As they grow they disperse downward on plants, feeding as they go. Larval development takes about 28 days. When fourth instar larvae have finished feeding, they wander or drop to the ground where they locate a suitable site to go underground to pupate. They use the surrounding soil to form their pupal capsules, preferring drier areas with a high relative humidity. They are in capsules 11 days to 42 days, but this stage averages about 28 weeks at 77°F (25°C). Total time from egg to adult is about 48 days. A female lays about 470 eggs in her lifetime.

Relative Effectiveness

The melaleuca weevil is being evaluated as a biocontrol agent of melaleuca in Florida following a widely publicized release ceremony at Ft. Lauderdale, on April 26, 1997. It is not expected that the release of one species will control the paperbark tree but will more likely slow its spread.

Both adult and larval feeding interfere with normal plant processes. This feeding will slow plant growth and make plants more susceptible to other control measures. Seed production may be reduced as well. As many as 200 larvae, which consume ten times more than adults, have been found on one melaleuca tree in Australia - a good omen for Florida.

Pesticide Susceptibility

Unknown.

MILE-A-MINUTE WEED CONTROL AGENT

Rhinoncomimus latipes (Coleoptera: Curculionidae)

Mile-a-minute weed, *Persicaria perfoliata* (Polygonaceae), is an annual vine that has been a pest in the eastern United States since its accidental introduction from Asia into York County, Pennsylvania, in the 1930s. It forms dense monocultures, scrambling over and suppressing other plants and preventing tree regeneration. The host-specific weevil, *Rhinoncomimus latipes*, occurs over a wide geographic range in China, Japan, Korea and the Russian Far East. Following intensive host-range testing, *R. latipes* was approved for release in North America in 2004. It was first released in Delaware and New Jersey, and has subsequently been released in nine other states.

Appearance

Adult weevils are about 2 mm long and are black when they first emerge, but become coated with an orange-brown substance apparently derived from plant sap after they have been feeding on *P. perfoliata* for several days. Weevils are typically found near the actively growing plant terminals. Feeding damage consists of characteristic small holes in leaves. Eggs are laid singly on the leaves, stems, or buds of *P. perfoliata*, and are yellowish, peanut-shaped, and covered with a small strip of frass. Larvae are rarely seen because they develop inside *P. perfoliata* stems, but scars or small holes are often visible on plants near nodes, where larvae have entered the stem or exited for pupation.

Habitat

Rhinoncomimus latipes is found only on *P. perfoliata*, both in its native and introduced range. *Persicaria perfoliata* is primarily riparian in its native range in Asia. In North America it grows in full sun or partial shade, with typical habitats including roadsides, edges of woods, reforestation clear-cuts, utility rights-of-way, open meadows (if unmowed), and stream banks.

Pests attacked

Host range studies in both China and North America have shown that this insect is extremely host-specific, with no egg-laying or larval survival on any plant species except *P. perfoliata*. Adult weevils feed minimally on several related Polygonaceae, but only under laboratory no-choice conditions.

Life cycle

Adult weevils overwinter in the soil or leaf litter. They emerge in early spring when *P. perfoliata* seedlings are about 6 inches tall, and begin mating and laying eggs. Newly hatched larvae bore into the plant stem at nodes and feed internally. Once they are full grown, larvae crawl or drop out of the stem and pupate in the soil under the host plant. Adult weevils emerge from the soil about one week later, crawl up nearby stems and mate on the host plant, feed on its leaves, and initiate egg-laying. The complete *R. latipes* life cycle from egg to adult takes about one month, and at least three to four overlapping generations occur during the growing season in the mid-Atlantic region. Egg laying ceases between mid-August and mid-September, as weevils prepare for overwintering.

Relative effectiveness

Although it is too soon for a complete assessment of the effectiveness of *R. latipes* in controlling mile-a-minute weed, initial indications are very positive. Weevils have overwintered and established at virtually all sites where they have been released, and *P. perfoliata* populations have declined substantially at many sites. Weevils disperse at a rate of more than 4 km per year after the first year of release. They locate both large and small mile-a-minute weed patches and establish new populations.

Pesticide susceptibility

This weevil is susceptible to commonly used insecticides and thus should not be exposed to insecticidal sprays. It is not directly affected by commonly used herbicides, but larval development is indirectly affected by herbicides if the plant dies before larval feeding is completed.

MUSK (NODDING) THISTLE AND PLUMELESS THISTLE CONTROL AGENTS

Rhinocyllus conicus (Coleoptera: Curculionidae)

Rhinocyllus conicus is a species of true weevil. It is best known as a controversial agent of biological pest control which has been used against noxious thistles in the genera *Carduus, Cirsium, Onopordum,* and *Silybum.*

The adult weevil is black and covered in a thin black and yellowish mottled coat of hairs. It is a short-snouted beetle up to 6 millimeters in total body length. The female lays over 100 eggs on or near the bracts of the thistle flower head. She covers the eggs with masticated plant tissue to protect them from predators. When the white larva emerges from its egg it burrows into the flower head and feeds on the flower parts and developing seeds. As it grows it deposits frass and chewed plant tissue on the walls of its chamber, producing a rigid protective shell in which it will pupate. Pupation takes up to two weeks and when the weevil emerges as an adult it remains inside the chamber for a few more weeks before tunneling out of the plant.

Damage to the plant occurs mainly from larval destruction of the flower head, which prevents seed production. Some larvae tunnel through the upper stem instead of chambering in a flower head; this can also be destructive to the plant. Adults do some damage as well when they feed on the foliage. Thistles which reproduce only via seed, such as musk thistle, are controlled well by this weevil and its seed head destroying larvae. Some thistles are able to reproduce vegetatively, and while they are impacted as well, they can sometimes survive.

This weevil is native to Eurasia and North Africa. It was first introduced to the United States for thistle biocontrol in 1969, and it is now widely established in that country. The weevil was found to be very effective in reducing the spread of invasive thistles, particularly musk thistle, but also welted, Italian, bull, milk, and other thistles. However, the weevil will also readily attack native thistles of genus *Cirsium,* in some cases contributing to population decline. It is for this reason that this weevil is no longer recommended for distribution as a biocontrol agent. Further releases of the weevil are prohibited in many areas.

Appearance

The adults are dark brown in color and about 10 to 15 mm long. In early spring, they are yellowish brown in color due to the yellow tufts of hair on their wings. As they age, they lose some of these hairs and turn brownish black.

Habitat

Musk thistle and plumeless thistles are introduced Eurasian noxious weeds in pastures, rangelands, croplands, and along state highways in many parts of the USA. Their success is largely due to their prolific seed production, seed longevity, competitive ability of the plants, and the lack of natural enemies.

Pests Attacked

This weevil is quite host specific and feeds mainly on thistles belonging to the *Carduus, Cirsium* and *Silybum* group in the family Asteraceae. None of the economical plants belonging to the same family is attacked by these weevils.

Life Cycle

R. conicus overwinters in the adult stage. It becomes active in mid to late April, depending on temperature. Each female lays about 100-200 eggs. Oviposition is well synchronized with musk thistle development. Eggs are laid on the bracts of developing buds and hatch in about 6 days.

Newly hatched larvae feed through the bracts into the buds. The larvae feed on the receptacle and prevent the production of viable seeds. They complete development in 4 to 6 weeks and turn into pupae in the thistle heads. The new adult emerges in 7-10 days. These new adults do not stay on the plants for long. They seek shelter and pass the summer in hiding, and then hibernate in winter. They re-emerge in spring to lay eggs before dying. There is one generation a year.

Relative Effectiveness

This thistle-head weevil is very effective in the control of musk or nodding thistle. The first success in biological control of musk thistle was documented in 1975 soon after the weevils were released in Virginia. Typical musk thistle stand reductions of 80-95% occurred in sites where the weevil became established. However, it only provides partial control of plumeless thistle because the egg laying period of the weevil only covers the development of the terminal thistle heads, but not the lateral heads.

Pesticide Susceptibility

This weevil is susceptible to the commonly used insecticides and thus should not be exposed to insecticidal sprays. It is not directly affected by commonly used herbicides and can be used in conjunction with herbicides for thistle control. Larval development is indirectly affected by herbicides if the thistle dies before larval feeding is completed. Herbicides are best applied after the insect has completed development.

Trichosirocalus horridus (Coleoptera: Curculionidae)

Trichosirocalus horridus is a native of south and central Europe that feeds on the rosettes of *Carduus* thistles. This rosette weevil was the second insect introduced into North America for *Carduus* thistle control. It was imported from Italy into the USA in 1970-1972 for quarantine testing. Host specificity testing revealed that it would not be a threat to nontarget plants and it was approved for field release in the USA in 1974. It first became established in Virginia and has since been relocated to many other states.

Appearance

The adult is dark brown to black in color and about 10 mm long. Newly emerged adults are reddish. They change to dark brown to black with time.

Habitat

Nodding (musk) thistle and plumeless thistles are introduced Eurasian noxious weeds in pastures, rangelands, croplands, and along state highways in many parts of the USA. Their success is largely due to their prolific seed production, seed longevity, competitive ability of the plants, and the lack of natural enemies.

Pests Attacked

This rosette weevil is host specific and feeds mainly on thistles belonging to genera *Carduus*, and *Cirsium* in the subtribe Carduinae of the Asteraceae family. It does not attack economical plants belonging to the Asteraceae.

Life Cycle

T. horridus overwinters in the adult, egg and larval stages. Adults emerge in late winter and early spring to lay eggs on young thistle rosettes. Each female lays an average of 500-800 eggs. Eggs are laid within the mid-rib of the underside of the leaves in clusters of 2 or 3, and hatch in about 10-12 days. Newly hatched larvae feed within the mid-rib towards the center growth point of the thistle rosette. The larvae reach the growth point in 7 days and continue feeding on the young tissues, causing the center point to become dark in color. They complete development in 6-8 wks. The full grown larva leaves the plant to pupate in the soil in pupation chambers. The new adult emerges in 12-20 days. The new generation of adults appears in May or June and goes into hiding until the fall, when they emerge to feed. Some may start laying eggs until the first frost before hibernation. There is usually one generation per year.

Relative Effectiveness

This thistle rosette weevil is very effective in the control of musk as well as plumeless thistle. The first success in biological control of musk thistle was documented in 1986. Biological control of musk thistle occurs after 5-6 years but it takes longer (10-12 years) to control plumeless thistles. Typical thistle stand reductions of up 95% were obtained at sites where the weevil became established. It also combines well with the thistle-head weevil as they attack different stages of the plant growth.

Pesticide Susceptibility

This weevil is susceptible to the commonly used insecticides and thus should not be exposed to insecticidal sprays. It is not directly affected by commonly used herbicides and thus can be used in conjunction with herbicides for thistle control. This can be carried out as long as the insect has completed development before the thistles die from herbicide action.

PURPLE LOOSESTRIFE CONTROL AGENTS

Galerucella calmariensis and *G. pusilla* (Coleoptera: Chrysomelidae)

Galerucella calmariensis is a species of leaf beetle in the family Chrysomelidae. It is commonly known as the black-margined loosestrife beetle and is native to Europe and Asia where both adults and larvae feed on purple loosestrife (*Lythrum salicaria*). It has been introduced in North America as a biological control agent for purple loosestrife.

Galerucella calmariensis

These native European beetles were introduced to North America in 1992 as part of a 5-15 year program to control purple loosestrife, an exotic weed infesting North American wetlands. Release sites were New York, Pennsylvania, Maryland, Virginia, Minnesota, Oregon, and Washington state in the United States, and sites in Canada. Since 1992, releases have been made in Ohio, Indiana, Illinois, Iowa, Michigan, Wisconsin, South Dakota, Colorado and Montana. Colonization of these introduced populations appears to have been successful.

Description

The black-margined loosestrife beetle is cylindrical in shape, mid brown, three to six millimetres long and half that width. When seen from above it often has two blackish lateral lines down either side. The eggs are spherical, white and usually topped with frass. The larvae are yellow speckled with black and resemble small caterpillars.

Biology

Adults overwinter in leaf litter near purple loosestrife plants. They emerge in the spring and start to feed on the young growth. They cause damage to the leaves characterised by neat rounded shot holes between the veins. The females lay up to 500 eggs from May to July in small batches on leaves and stems. On hatching, the first instar larvae feed on growing shoot tips and may kill them. Later instars feed on stems and leaves leaving the upper cuticle of the leaf intact. When the density of the larvae is high enough, the plant can be completely defoliated. After feeding for about three weeks and undergoing further moults, the larvae move down the plant to pupate in the soil or leaf litter. If the water level is high, larvae sometimes pupate in the hollow aerenchyma tissue in the plant stems. Both adults and larvae can float and may be wafted by water currents or wind to new host plants.

Appearance

G. calmariensis: Light brown, usually with a black triangle or a broad, dark stripe on the thorax. The body has parallel sides, is 3-5 mm long and half as wide, and is punctuated with coarse dots and fine, dense hairs.

G. pusilla: Light brown, usually with a narrow dark stripe on the thorax. The body has fine, dense hairs, parallel sides, is 3-4 mm long and about half as wide, with a slight narrowing toward the head. *G. pusilla* is also punctate, but less coarsely so than *G. calmariensis*.

Habitat

Purple loosestrife is a weed species in wetlands over much of temperate North America, and the aim of the introduction program of *G. calmariensis* and *G. pusilla* is to locate the releases so that these beetles will be able to easily colonize and spread. Currently, purple loosestrife exists in monotypic stands throughout the northeastern U.S. and southeastern Canada, the Midwestern U.S., and in isolated locales in the western U.S. and southwestern Canada. Irrigation systems provide conduits for its spread in dry areas.

Pests Attacked

G. calmariensis and *G. pusilla* are considered host-specific to purple loosestrife (*Lythrum salicaria*). Before introduction to North America, approximately 50 native plants, including some close relatives of purple loosestrife, were tested for susceptibility to these beetles. Only winged loosestrife (*Lythrum alatum*) was a potential host, and under field conditions in Europe, it was determined that if given a choice, *G. calmariensis* and *G. pusilla* avoided the North American native.

Life Cycle

G. calmariensis and *G. pusilla* have very similar life habits. Adults emerge in spring from hibernation in leaf litter and feed on the new leaves and shoots of purple loosestrife. The egg laying phase lasts approximately two months in the spring and eggs are laid in clusters of two to ten daily on the plant stem and in the leaf axils. A female can lay 300-400 eggs per year, and the adult lives 8-10 weeks. Larvae feed on bud, leaf, and stem tissue. Pupation takes place in the soil or ground cover near the plants. From egg to adult takes about 6 weeks, and there is generally one generation per year.

Adults are very mobile and successful in seeking out new stands of purple loosestrife. Most searching for a host plant occurs after hibernation when plant growth is at a peak, and again following the emergence of new beetles in July and August. Once a host has been located, migration slows. The overwintered adults die by late June, soon after the reproduction phase has been completed.

Relative Effectiveness

In Europe, there are several insects which act together to control purple loosestrife. As a result, only small, scattered stands of the plant exist. The current program includes the introduction of these *Galerucella* spp., the weevil *Hylobius transversovittatus* and the planned release of two flower-eating species which are also specific to purple loosestrife. It is predicted that upon establishment of these species, North American purple loosestrife will be reduced by 90% over approximately 90% of its present range.

Plants growing in high water and in shade are not hospitable sites for *Galerucella*.

Pesticide Susceptibility

Early indications are that the *Galerucella* spp are very susceptible to pesticides; exposure should be strictly avoided. Conservation sites where purple loosestrife is continuously distributed and which are free of standing water most of the year are optimal. To establish a colony of *Galerucella* will probably take 7-10 years and the site (5-10 acres is adequate) should be safe from land development and broad spectrum insecticides during that time.

Use in biological control

Purple loosestrife is a native of Europe, Asia, northwest Africa and parts of Australia. In these countries a number of insects are associated with it. These include the weevil *Hylobius transversovittatus* and the leaf beetles *G. calmariensis* and *G. pusilla*. When purple loosestrife was introduced into North America no insects were available to keep it under control and it became rampant. It is now found in large stands throughout the northeastern United States, southeastern Canada and the American Midwest where it crowds out other species and is a threat to biodiversity.

The black-margined loosestrife beetle and the closely related golden loosestrife beetle (*Galerucella pusilla*) have been released in over 27 states in the United States and 6 Canadian provinces. The adults of both species have become well established and proved very effective. They show great ability to find new stands of the host plant and can locate patches of purple loosestrife a kilometre away within a few days of emergence.

Before these beetles were introduced to North America, the susceptibility of approximately 50 native plants of North America to the beetles was tested in Europe. Among the plants were several close relatives of purple loosestrife. Only winged loosestrife (*Lythrum alatum*) proved to be susceptible, and was not used by the beetles as a host when purple loosestrife was available.

Hylobius transversovittatus (Coleoptera: Curculionidae)

Hylobius transversovittatus is a species of weevil in the family Curculionidae. It is native to the Old World where both adults and larvae feed on purple loosestrife (*Lythrum salicaria*). This plant is regarded as an invasive species in North America and the weevil has been introduced into both the United States and Canada in an effort to control the plant.

This native European weevil was introduced in 1992 as part of a 5-15 year program to control purple loosestrife, an exotic weed infesting North American wetlands. Release sites were New York, Pennsylvania, Maryland, Virginia, Minnesota, Oregon, Washington state in the United States, and sites in Canada. Since 1992, releases have been made in Ohio, Indiana, Illinois, Iowa, Michigan, Wisconsin, South Dakota, Colorado and Montana. Colonization of these introduced populations appears to have been successful.

Description

The adult *Hylobius transversovittatus* is a dark brown colour with two irregular transverse lines of tufts of white hair. It is about thirteen millimetres long and six millimetres wide and has a narrow head and thorax and a curved trunk-like proboscis. The antennae are elbowed and the legs reddish. [3] The eggs are white or pale yellow and oval in shape and hatch within about two weeks. The larvae are a creamy colour with dark brown heads and adopt a crescent shape.

Biology

Adult *Hylobius transversovittatus* beetles emerge in the spring from the soil or leaf litter where they have been overwintering. They feed on the leaves of purple loosestrife (*Lythrum salicaria*) and are mostly nocturnal. Their presence is evidenced by the ragged edges to leaves caused by their chewing. They are most active from June to August and may live for several years. The females lay about 200 eggs over a period of two to three months with each being deposited singly near the roots of purple loosestrife or occasionally on the stem. On hatching, the soil-laid larvae feed on root hairs before burrowing their way into the root where they feed on tissue within the woody rootstock. Stem-laid larvae tunnel into the stem and work their way down to the root. Larval development may be interrupted by periods of flooding and resume when the water level falls. The larvae undergo two moults over a period of one to two years before forming pupation chambers in the upper part of the root. They then moult again and become pupae. When metamorphosis is complete, the adult beetles chew their way out, usually emerging between July and October. The weevils can overwinter in any of their life stages, as eggs, larvae, pupae or adults.

Appearance

The *Hylobius transversovittatus* adult is about 13 mm long and 6 mm wide, is dark brown and has two rows of white tufted hairs across its body. Its thorax and head taper to an elongate, curved proboscis. Eggs are oval and whitish, about 1.4 mm long and 1 mm wide.

Habitat

Purple loosestrife is a weed species in wetlands over much of temperate North America, and the aim of the introduction program of *H. transversovittatus* is to locate the releases so that this weevil will be able to easily colonize and spread. Currently, purple loosestrife exists in large, monotypic stands throughout the northeastern U.S. and southeastern Canada, the Midwestern U.S., and in isolated locales in the western U.S. and southwestern Canada. Irrigation systems provide conduits for its spread in dry areas.

Pests Attacked

H. transversovittatus is considered host-specific to purple loosestrife (*Lythrum salicaria*). Before introduction of *H. transversovittatus* to North America, approximately 50 native plants, including some closely related to purple loosestrife, were tested for susceptibility to this weevil. Only two, swamp loosestrife (*Decodon verticillatus*) and winged loosestrife (*Lythrum alatum*), were potential hosts, and under field conditions in Europe, it was determined that *H. transversovittatus*, if given a choice, avoided the North American natives.

Life Cycle

Adults emerge in the spring from soil or soil litter and feed at night on the new leaves of purple loosestrife. Evidence of their feeding is seen along leaf edges. Peak activity is from June to August, although their reclusive, nocturnal patterns make them difficult to observe. Adults may live for several years. The egg laying phase lasts two to three months, with one to three eggs being laid each day and deposited singly into the stem or the soil near a root. A female can produce approximately 200 eggs each year. The larvae mine to the roots where they feed on root tissue. Larvae develop in 1-2 years. Mature larvae form a pupation chamber in the upper part of the root from which they emerge as adults between July and October. Larvae hibernate in different stages of maturity.

Relative Effectiveness

In Europe, there are several insects which act together to control purple loosestrife. As a result, only small, scattered stands of the plant exist. The current program includes the introduction of this *Hylobius* species, two species of the beetle *Galerucella,* and the planned release of two flower-feeding beetle species which are also specific to purple loosestrife. It is predicted that upon establishment of these species, North American purple loosestrife will be reduced by 90% over approximately 90% of its present range.

These weevils will not prosper on plants growing in high water areas and in shade.

Pesticide Susceptibility

Early indications are that *H. transversovittatus* is highly susceptible to pesticides; exposure should be strictly avoided.

Conservation

Although beetles and larvae can survive submergence for a few days or weeks, experiments have shown that excessive flooding will eventually prove fatal to larvae and restrict adult access to the plant. Best sites should be at least 5-10 acres and free of standing water most of the year. The optimum condition is ground water 1-3 inches below the surface.

Use in biological control

Purple loosestrife is a native of Europe, Asia, northwest Africa and parts of Australia. In these countries a number of insects are associated with it. These include *Hylobius transversovittatus,* the adults of which eat the foliage and the larvae of which tunnel through and eat the root. When purple loosestrife was introduced into North America these insects were not available to keep it under control and it became rampant. It is now found in large stands throughout the northeastern United States, southeastern Canada and the American Midwest where it crowds out other species and is a threat to biodiversity.

The weevils and their eggs were introduced to test sites in Virginia from 1992 onwards. They became successfully established and six years after the introduction, 28% of the roots of purple loosestrife plants were found to be infested. Larvae were also detected in roots four hundred metres downstream from the release point. Although the feeding of the adult weevils does little damage to the plant, the activities of the larvae, especially if there are several in one rootstock, weaken the plant, make it more susceptible to stressful conditions and reduce the quantity of seed produced

ST. JOHNSWORT CONTROL AGENT

Chrysolina quadrigemina (Coleoptera: Chrysomelidae)

KLAMATHWEED BEETLE

The Klamathweed beetle, Chrysolina quadrigemina, is native to northern and central Europe and western Asia. The first North American introduction was reported from California in 1945 where it was introduced as a biological control for St. John's Wort (Hypericum perforatum) which was listed on the federal noxious weed list. The beetle has become well established in California, Idaho, Montana, Oregon, Washington and most recently, western North Carolina where it defoliates H. perforatum by targeting hypericin containing glands. The presence of the Klamathweed beetle in North Carolina was first reported in 1997 from a shipment of fresh St. John's Wort harvested from naturalized populations in Oregon. The beetles overwintered successfully and established a population in Brevard, NC. Additional populations were detected in 1999 from experimental plantings located in Ashe County, NC, at the Upper Mountain Research Extension Station and the Mountain Horticultural Crops Research & Extension Center (MHCREC) in Fletcher, NC.

The recent popularity of St. John's Wort (H. perforatum) for its reported antidepressant and antiviral properties has prompted interest in commercial production. Because commercial demand exceeded available supply of the crop in 1997, there is increased interest in propagation and associated pathogens.

Destructive defoliating stages of the beetle are both larval and adult stages. They are reported to be host specific to St. John's Wort (H. perforatum) and defoliate plants in spring, summer, and fall thus reducing the plants' ability to overwinter successfully.

Defoliation is the main damage produced by the beetle and does not normally exceed 25-50%, thus rarely causing mortality. The most damaging result is a reduction in commercial quality as the beetle specifically targets hypericin producing glands along petal margins. However, when the fungal pathogen Colletotrichum gloeosporioides (Anthracnose) is present in the soil or on the plant, the beetle becomes a vector for the spores, causing mortality rates to increase dramatically from 60-90%.

The beetle produces one generation per year and overwinters in the egg, larval, and occasionally adult stage. Eggs are laid singly or in clusters on the underside of foliage in fall or spring. Eggs laid in fall survive winter conditions and hatch the following spring. An individual female may lay thousands of oval, orange to reddish eggs per generation. Eggs laid in spring will hatch after approximately three weeks when larvae migrate to flower buds and immature leaves. Larvae feed during early A.M. then return to soil for the remainder of the day. Young larvae are C-shaped and orange becoming gray-pink with age. Larvae continue to foliage feed until reaching maturity then burrow into soil where they create cells and pupate in late February and March. They emerge as oval orange beetles with wing pads, legs, head, and antennae visible in late April until June. Adults are attracted to terminal branches toward sunlight and feed during the warmest period of the day.

Adults emerge in spring and feed for several weeks before returning to soil for a resting stage. Fall rains activate the adults to mate and lay eggs. Adults are metallic, bronze, black, green, oval beetles 5-7 mm (0.2-0.28 in) long. Chrysolina sp. are invariably sedentary but will fly in response to decreases in their host plants or high temperatures (40°C). Studies have proven that the beetles can move freely to widely separated populations of St. John's Wort (*H. perforatum*).

Appearance

C. quadrigemina is up to 1/4" in diameter. Although not brightly metallic, it has a metallic sheen that can be green, bronze, purple, and/or blue, depending on the individual.

Habitat

Rangeland infested with *Hypericum perforatum*.

Pests attacked

C. quadrigemina is generally specific to *H. perforatum*. It has been reported feeding on some native and ornamental Hypericum species, but it does not prosper. It is not known if it can complete its life cycle on species other than *H. perforatum*.

The Klamathweed beetle is very closely synchronized in its growth stages with the host plant. In 1943, negotiations between Harry S. Smith, Chairman of the Department of Biological Control at the of the University of California and the USDA Bureau of Entomology and Plant Quarantine, resulted in an agreement that *C. quadrigemina* and two other potential Klamath weed control agents would be tested, prior to release, on 6 plants to which the insects had not been exposed in Europe or Australia: sugar beet, flax, hemp, sweet potato, tobacco, and cotton. Upon satisfactory completion of the tests, the Klamathweed beetle was released.

Life cycle

Adults emerge from pupal cells that are just beneath the soil surface in April and early May when the plants are producing flower buds or are actually in flower. They feed voraciously on the plant foliage until late June/early July, and then, fully fed, spend an obligate period of aestivation beneath stones and debris and in crevices of the soil. The aestivation period cannot be entered in moist conditions. With the onset of fall rains, the beetles again become active, in synchrony with the host plants. The plants form a leafy, basal growth upon which active beetles feed sparingly, then mate and lay their eggs about mid-October. The larvae reach the third- and fourth-stages of development by feeding on the plant's basal growth, which they completely destroy. They then enter the soil to a depth of about one inch and individually form cells for pupation. There is one generation per year.

Relative effectiveness

Because its life cycle is so well synchronized with Klamath weed growth habits, *C. quadrigemina* is very effective against this plant pest if it is located in climates suitable for the beetle.

After the initial North American release in 1946, *C. quadrigemina* multiplied so quickly that at one location two years later, 5000 beetles, a small fraction of the total population, were collected for redistribution. By 1951, populations were established in all 21 California counties in which Klamath weed was a problem, and the plant was already under control in some of the original locations. By 1957, the plant was reduced by 99% in Northern California.

C. quadrigemina is adapted to a climate with hot, dry summers and mild, rainy winters, and all stages of *C. quadrigemina* – eggs, larvae, pupae, and adults – are able to withstand winter temperatures of California areas that support Klamath weed. Adults and larvae are susceptible to frosts, but all stages can survive under snow cover.

Canadian efforts have had mixed success. In 1951, releases in British Columbia of *C. quadrigemina* and *C. hyperici* (a species more tolerant of moisture) have resulted in effective control of St. Johns wort in many areas of the province. Both of these species have been introduced and become established in southern Ontario. More recently, *C. quadrigemina* has been reported in New York, Pennsylvania, Ohio, Maryland, and West Virginia, and is presumed to have spread from release sites in Ontario.

In areas where the Klamathweed beetle thrives, St. Johns wort is reduced to an acceptable level or is at least subjected to extensive damage. However, in moist areas, *C. quadrigemina* introductions have done little to control the weed, even when the beetle persists. Cold is another negative factor, and in one year of heavy frost, the *C. quadrigemina* population at a Canadian site (Zamora-Westbridge vicinity) collapsed. The presence of trees and dense ground vegetation helps to protect the beetles from frosts.

Conservation

The conservation of natural enemies is probably the most important and readily available biological control practice available to growers. Natural enemies occur in all production systems, from the backyard garden to the commercial field. They are adapted to the local environment and to the target pest, and their conservation is generally simple and cost-effective. With relatively little effort the activity of these natural enemies can be observed. Lacewings, lady beetles, hover fly larvae, and parasitized aphid mummies are almost always present in aphid colonies. Fungus-infected adult flies are often common following periods of high humidity. These natural controls are important and need to be conserved and considered when making pest management decisions. In many instances the importance of natural enemies has not been adequately studied or does not become apparent until insecticide use is stopped or reduced. Often the best we can do is to recognize that these factors are present and minimize negative impacts on them. If an insecticide is needed, every effort should be made to use a selective material in a selective manner.

TROPICAL SODA APPLE

Gratiana boliviana (Coleoptera: Chrysomelidae)

Tropical soda apple (TSA), *Solanum viarum* Dunal (Solanaceae), is a perennial weed, native to Brazil, Argentina, Paraguay, and Uruguay, that has been spread throughout Florida very rapidly during the last two decades. TSA was first reported in Glades County in 1988. This weed is also present in Alabama, Arkansas, Georgia, Louisiana, Mississippi, North Carolina, Pennsylvania, South Carolina, Tennessee, and Texas . Currently, the area infested with TSA is estimated at more than one million acres.

TSA invasion is a major problem for ranchers due to the loss of pasture area and subsequent lower stocking rates. TSA is a perennial shrub, approximately 1 to 1.4 m tall, and covered with prickles and tiny hairs that contain a sticky substance. TSA fruits are 2 -3 cm in diameter and green when immature with a watermelon color pattern, and turn yellow when ripe. Cattle and small mammals are known to feed on TSA fruits, thus, facilitating the spread of seeds to new habitats. The dominance of TSA is most noticeable during the summer and fall when plants are actively growing due to high temperatures and adequate moisture. Freezing winter temperatures kills TSA above the ground, but new shoots protected by leaf litter and roots often survive to resprout when warmer weather arrives.

Because of the importance of TSA as an invasive weed, a search for biological control agents was initiated in South America in 1994. Several insect herbivores were found causing damage to TSA in the native range. One agent identified and later introduced from Paraguay and northern Argentina was the TSA leaf feeding beetle, *Gratiana boliviana*.

First Biocontrol Agent for Tropical Soda Apple

The TSA tortoise beetle, *Gratiana boliviana* Spaeth (Coleoptera: Chrysomelidae) was approved by the USDA-APHIS-PPQ for field release in Florida on May 7, 2003. The initial release of *G. boliviana* in Florida began in May 14 in Polk County. Since the summer of 2003, approximately 233,000 beetles have been released in Florida, Georgia, Alabama, and Texas. This beetle has been established and is dispersing at all the release sites in south/central Florida and Jasper County, Texas.

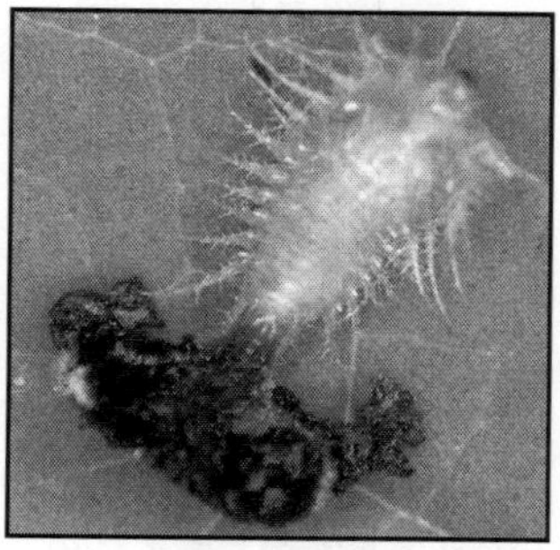

Gratiana boliviana **larva**

Gratiana boliviana lays individual eggs on TSA leaves and petioles. Eggs are initially white in color but turn light green during the incubation period. Each egg is enclosed by two translucent brown membranes. The egg case is attached to the leaf surface by one extreme. A female can produce on average 300 eggs during her 3 to 4 month life cycle. Incubation of the egg takes 5-6 days at a temperature of 25°C (77°F).

Larvae are cream with a small green spot in the anterior half. Each segment of the body has two lateral processes and a pair of long caudal processes or small forks that arise from the last abdominal segment. Like most tortoise beetles, larvae carry the cast skins on the anal forks with the posterior end of the larva bent forward. The five larval instars can be completed in 15-18 days. Larval feeding is concentrated mostly in the upper third of the plant canopy. Infested plants are easily detected by clusters of small to medium size holes made by the feeding larvae. As feeding progresses, the plants may exhibit large areas of defoliated leaves. Almost complete defoliation has been observed in enclosed cages when the beetle's natural enemies are excluded.

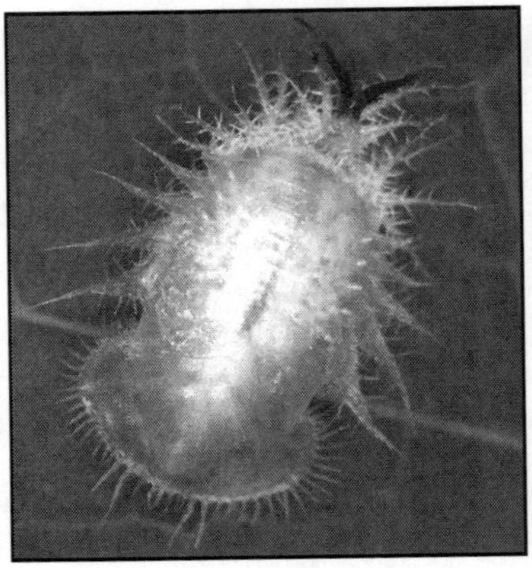

Gratiana boliviana **pupa**

The pupal stage is completed in about 6-7 days. The pupae are green and dorsal-ventrally (from top and bottom) flattened. Pupae are attached to the leaf by the last abdominal segment. The most common pupation place is the underside of leaves. In severely defoliated plants, some pupae can be found on the petioles and stems.

General coloration of the young adult is light green. Along the margin of the elytra (front wings) there is a continous yellow band. The rest of the elytra is light green with irregular yellowish areas between rows or depressions. Mature adults turn a uniform yellow. Females and males can be distinguished by examining the underside of the body. In males, two somewhat rounded orange testes can be observed, one on each side of abdomen. In females there is a pair of white ovaries. Characters for separating the sexes can be observed 3-5 days after emergence. Pre-oviposition (pre-egg laying) period takes from 9 to 12 days. Longevity of females averages 3 months.

Appearance

Gratiana boliviana has four developmental stages (egg, larva, pupa, adult): Eggs are brown and enclosed in a membraneous envelope, *larvae* are spiny and are pale green, *pupae* are spiny, flattened and immobile; and *adults*, are about ¼″ long and nearly as wide, and are a deep green color. Adults and larvae usually feed on the upper side TSA leaves while pupae and eggs can be found on the underside of leaves.

Habitat

Tropical soda apple-infested grasslands and wooded areas.

Pests attacked

Scientists from University of Florida and US Department of Agriculture exposed 123 plant species in 35 plant families to *G. boliviana* and evaluated beetle feeding and survival. *Gratiana boliviana* completed development only on TSA. On eggplant, very minor feeding damage was seen in laboratory tests, but none in the field. Biological control agents released to control invasive weeds are studied for several years in order to confirm that they will only feed on the target weed, in this case TSA.

Life cycle

The life cycle of the beetle begins when a female lays an egg. Eggs hatch in 5-6 days, larval development takes about 16-18 days, and the pupal stage lasts 6-7 days. Therefore, from egg to adult the beetle takes about 29-31 days to complete development at 77° F. Like all insects, development occurs faster when it's warm, and slower when it's cool. The beetles actively feed and reproduce in the field from around March/April until October/November, which is probably sufficient time to complete 7-8 generations. However, for 4-5 months during the winter, the beetles enter an adult resting state called 'diapause'. During diapause, the beetles feed very little, and do not reproduce. They are difficult to find at this time of the year, as they hide in leaf litter beneath plants.

Relative effectiveness

Larvae and adults of the beetles feed on TSA leaves, with a preference for feeding on younger leaves near the top of plants. The leaf feeding damage can be found from April to November and it is recognized by the characteristic "shot-gun" holes. This damage debilitates TSA and creates wounds that may facilitate attack by plant diseases. The cumulative stress produced by the beetles and diseases reduces the plant size and fruit production. Once TSA becomes less competitive, pasture grasses and native vegetation recover.

Studies conducted by scientists at the University of Florida have demonstrated that *G. boliviana* can reduce the density of TSA in as little as one year from the time of release, although in some cases it may take longer. *Gratiana boliviana* will not completely eliminate the weed from an infested area. Once the density of TSA is reduced, the beetle population will decrease due to a lack of food. Eventually, TSA and beetles will reach an equilibrium where the density of beetles regulates the density of plants, and the density of plants regulates the number of beetles.

Pesticide susceptibility

Beetles are in diapause from about November to April. The use of herbicides or mowing during this period will have minimal effect on beetles which are in leaf litter under plants.

Conservation

The conservation of natural enemies is probably the most important and readily available biological control practice available to growers. Natural enemies occur in all production systems, from the backyard garden to the commercial field. They are adapted to the local environment and to the target pest, and their conservation is generally simple and cost-effective. With

relatively little effort the activity of these natural enemies can be observed. Lacewings, lady beetles, hover fly larvae, and parasitized aphid mummies are almost always present in aphid colonies. Fungus-infected adult flies are often common following periods of high humidity. These natural controls are important and need to be conserved and considered when making pest management decisions. In many instances the importance of natural enemies has not been adequately studied or does not become apparent until insecticide use is stopped or reduced. Often the best we can do is to recognize that these factors are present and minimize negative impacts on them. If an insecticide is needed, every effort should be made to use a selective material in a selective manner.

WATER HYACINTH CONTROL AGENTS

Neochetina bruchi and *Neochetina eichhorniae* (Coleoptera: Curculionidae)

The mottled water hyacinth weevil, *Neochetina eichhorniae,* is a beetle insect that has been introduced as a biological pest control herbivore agent to waterways and lakes in countries worldwide to control the spread of the invasive noxious weed species *Eichhornia crassipes* - water hyacinth, an aquatic plant native to the Amazon basin.

Neochetina eichhorniae is native to Argentina and neighboring areas of South America. It is primarily a brownish-gray with characteristic brown mottling. It measures approximately 3.5 mm in length, excluding the head and antennae.

The weevil's life cycle is approximately three to four months, depending on environmental factors.

Neochetina eichhorniae and *N. bruchi* were introduced to control the exotic pest, waterhyacinth. Florida's sub-tropical climate and warm, shallow, fertile waters are ideal for waterhyacinth proliferation. This floating macrophyte was introduced into the United States at a New Orleans exposition in 1884. Samples are said to have been given to fair-goers, and within 4 years, coastal fresh waters were infested from Texas to Alabama. By 1972, the infestation in Florida was estimated to be 200,000 acres. Large, floating mats of waterhyacinth obstruct navigation, clog irrigation works, disrupt the natural ecology of wetlands in many ways, exacerbate mosquito problems, and are costly to the tourism and recreation industries.

Three insects were introduced into Florida in an attempt to control waterhyacinth. The weevils *N. eichhorniae* and *N. bruchi,* natives of Argentina and surrounding areas in South America, were released in 1972 and1974, respectively. The pyralid moth, *Sameodes albiguttalis,* was released in 1977. *N. eichhorniae* has proven to be the most important control agent.

Biological plant pest control

Neochetina eichhorniae was first introduced, as an herbivore agent for the biological control of water hyacinths, to Florida in the United States in 1972. Due to its successful reduction and control of the plant in the Everglades and other natural water systems there, it has subsequently been introduced in more U.S. states and other countries, such as in east African Lake Victoria as a powerful biological control agent.

Water hyacinth is a highly invasive species and has a tendency to cover and choke major waterways and lake surfaces, which can have numerous detrimental ecological, biological diversity, fisheries, hydroelectric, transportation, and economic results.

The adults produce characteristic feeding scars on the leaves and petioles. In the larval stage, the insect tunnels into the petioles and the crown of the plant. This feeding results in biotic stress, reduced flowers and seeds, and less vigorous growth.

Appearance

N. bruchi: The body is broad, robust, and densely covered with fused brown and tan scales. The tan scales form a V-shaped chevron on the back that distinguishes this species from *N. eichorniae*. Antennae and lower-leg segments are reddish-brown. There are yellowish water-shedding scales on the leg joints and parts of the underside. The snout is thick and weakly curved to straight in males; in females it is longer, more slender, and more curved. Male length is about 3.5 mm long (excluding head), and female length is about 4.5 mm long (excluding head). The female snout is noticeably shiny near the tip where the scales have been rubbed away. This easily distinguishes females from males in both species.

N. eichhorniae: The body is covered mostly with gray scales mottled with brown, and the antennae and lower-leg segments are reddish-brown. The snout is thick and weakly curved in males, and moderately slender and rather strongly curved in females. There are dense, yellow, water-shedding scales at the base of the head. The male is about 3.2 mm long (excluding head) and the female is about 3.7 mm long (excluding head).

Habitat

Waterhyacinth.

Pests Attacked

Waterhyacinth.

Life Cycle

The biologies of *N. bruchi* and *N. eichhorniae* are very similar. Studies in the United States show that adults are reproductive all year, without producing distinct generations. In Florida, females undergo a reproductive quiescence during the winter. Reproduction begins again after December. Oviposition is in the leaf tissue, *N. eichhorniae* placing eggs singly into holes chewed by the female in the leaf epidermis, and *N. bruchi* inserting eggs either singly or in groups of up to 25 into the petiole. About 30°C is optimal for feeding and oviposition in both species.

Eggs hatch in 7-17 days, depending on temperature. There are three instars, the larval stages taking from 36 to 90 days— reports vary widely. Larvae feed and develop in the petioles. They burrow downwards, usually arriving at the petiole base by the third instar. They may gouge pockets into the stem (rhizome) and feed on developing axillary buds. Fully grown larvae move into the water to the upper root zone, create a ball around themselves formed from lateral roots and pupate attached to the main root axis, possibly using the root as an oxygen source. Pupation lasts approximately 7-9 days at 27° C. Feeding begins within 24 hours of leaving the pupal chamber, and eggs are usually laid about 6 days later. Adults are nocturnal.

Total generation time is from 70 to 140 days for *N. eichhorniae* and 56 to 94 days for *N. bruchi*, depending on time of year. Females lay about 200 eggs during their lifetime and sometimes as many as 400 or more. Fecundity is heavily dependent on abundant, healthy plants.

Relative Effectiveness

The biological control of waterhyacinth is considered a qualified success. Waterhyacinth infestation in Florida was estimated to be only 2600 acres in 1989. Although this figure was obtained after herbicide treatment of 25,000 acres, earlier herbicide treatment of more acreage had not reduced the infested area. Also, many of the early study sites used to investigate the performance of waterhyacinth biological control agents were situated in undisturbed backwater locations, and most of those sites no longer harbor waterhyacinth. Although declines of these waterhyacinth populations were well documented and readily attributable to biological control, the documentation was never published.

Herbicide Tolerance

Although immature weevils have no way to escape dying and sinking plants, experiments have shown that *Neochetina adults are able to tolerate applications of 2,4-D, Kover II, diquat, and polymer "573" (see Haag, below, for concentrations) and can move freely between sprayed and unsprayed plants. It is therefore desirable to keep an area of waterhyacinth unsprayed in order to provide habitat for the weevil population needed to continue biological control.*

Conservation

The conservation of natural enemies is probably the most important and readily available biological control practice available to growers. Natural enemies occur in all production systems, from the backyard garden to the commercial field. They are adapted to the local environment and to the target pest, and their conservation is generally simple and cost-effective. With relatively little effort the activity of these natural enemies can be observed. Lacewings, lady beetles, hover fly larvae, and parasitized aphid mummies are almost always present in aphid colonies. Fungus-infected adult flies are often common following periods of high humidity. These natural controls are important and need to be conserved and considered when making pest management decisions. In many instances the importance of natural enemies has not been adequately studied or does not become apparent until insecticide use is stopped or reduced. Often the best we can do is to recognize that these factors are present and minimize negative impacts on them. If an insecticide is needed, every effort should be made to use a selective material in a selective manner.

Glossary

Abdomen: The last of the three major body divisions of an insect.

Action threshold: The pest density at which a control tactic must be implemented to avoid an economic loss.

Active ingredient (AI): The component of a pesticide formulation responsible for the toxic effect.

Agroecosystem: A relatively artificial ecosystem in an agricultural field, pasture, or orchard.

Antenna, Antennae (pl.): A pair of sensory organs located on the head of an insect, above the mouthparts.

Aorta: The front-most, non-pulsating portion of the dorsal blood vessel of an insect.

Arthropod: Any of the invertebrate animals (such as insects, spiders, or crustaceans) having an exoskeleton, a segmented body and jointed limbs.

Augmentation: Biological control practices intended to increase the number or effectiveness of existing natural enemies.

Bacterium: A single-celled microscopic plant-like organism that does not produce chlorophyll.

Beak: Colloquial expression for the protruding mouthpart structures of a sucking insect (= proboscis).

Biological control: The use of living organisms, such as predators, parasitoids, and pathogens, to control pest insects, weeds, or diseases. Typically involves some human activity.

Biorational: Having a minimal disruptive influence upon the environment and its inhabitants (e.g., a biorational insecticide). Broad-spectrum (insecticide): Active against a wide range of insects.

Bt: The bacterium, Bacillus thuringiensis.

Caterpillar: The immature stage (larva) of a butterfly, moth, or sawfly.

Chemical control: Pest management practices which rely upon the application of synthetic or naturally-derived pesticides.

Class: A category of the classification scheme of living organisms ranking below a phylum and above an order (e.g., Insecta).

Classical biological control: The importation of foreign natural enemies to control previously introduced, or native, pests.

Cocoon: A silken case formed by an insect larva for pupation.

Cole crops: Crops such as cabbage, broccoli, Brussels sprouts, and other crucifers.

Complete metamorphosis: Type of insect development characterized by four distinct stages: egg, larva, pupa, and adult.

Conservation: Any biological control practice designed to protect and maintain populations of existing natural enemies.

Contact poison: A pesticide that is absorbed through the body wall, as opposed to one that must be ingested.

Cucurbits: Vine crops such as cucumbers, melons, squash, and pumpkins.

Cultural control: Pest management practices that rely upon manipulation of the cropping environment (e.g., cultivation of weeds harboring insect pests).

Density (insect populations): The number of insects per unit of measure (e.g., beetles per square meter).

Diapause: A physiological state of arrested metabolism, growth, and development that occurs at a particular stage in the life cycle of an organism.

Dormancy: A recurring period in the life cycle of an organism when growth, development, and reproduction are suppressed.

Ecology: The study of an organism's interrelationship with its environment.

Economic threshold: see Action threshold.

Elytra: The thickened or leathery front or forewings of insects such as beetles.

Encapsulation: The surrounding of an invading body, such as the egg of a parasite, by insect hemocytes (blood cells) and the formation of a protective capsule.

Entomopathogenic: Insect-attacking organism.

Environmental impact quotient (EIQ): A relative value that estimates the environmental impact of a pesticide, by taking into account toxicity to natural enemies, wildlife, and humans, degree of exposure, aquatic and terrestrial effects, soil chemistry, etc.

Epizootic: A disease outbreak within an insect population.

Exoskeleton: A skeleton or supportive structure on the outside of an insect body.

Exotic: Introduced from another country or continent (e.g., introduced insect pest).

Family: A taxonomic subdivision of an order, containing a group of related genera. Family names end in -idae.

Forewing: The first or anterior pair of insect wings.

Fumigant: A substance which produces a gas, vapor, fume, or smoke intended to kill a pest.

Fungicide: Any substance that kills or inhibits the growth of a fungus.

Funicle: The portion of the flagellum of the antenna closest to the club.

Fungus, Fungi (pl.): Any of numerous plants lacking chlorophyll, ranging in form from a single cell to a body of branched filaments. Includes the yeasts, molds, smuts, and mushrooms.

Gene: A biochemical unit of hereditary, often coding for an entire protein.

Generalist: A pest or natural enemy that can utilize a wide range of species as host or prey.

Generation: Period from any given stage in the life cycle to the same life stage in the offspring. Typically from egg to egg.

Genetic engineering: The manipulation of the genetic material of an organism in order to achieve desirable characteristics.

Genus, Genera (pl.): A group of evolutionarily related species, sharing one or a number of characteristics.

Gradual metamorphosis: A type of insect development in which there is no prolonged resting stage (pupa). The three stages are: egg, nymph, and adult.

GV: Granulosis virus.

Habitat manipulation: Manipulation of agricultural areas and surrounding environment with the aim of conserving or augmenting populations of natural enemies (e.g., the planting of a refuge for natural enemies).

Head: The anterior region of an insect, which bears the mouthparts, eyes, antennae and houses the brain.

Herbicide: A substance used to kill or control weeds.

Hermaphroditic: Having both male and female sex organs in one individual.

Hindwings: The second pair of wings of an insect.

Honeydew: The sugary liquid discharge from the anus of certain insects (Homoptera) such as aphids and scales.

Host: The organism in or on which a parasitoid lives; a plant on which an insect feeds.

Host plant resistance: The relative amount of heritable qualities possessed by a plant that reduces the degree of damage to the plant by a pest or pests.

Hyperparasite: A parasite whose host is another parasite.

Indigenous: Native to an area.

Inoculative release: The release of relatively small numbers of natural enemies that are expected to colonize, reproduce, and spread naturally throughout an area.

Insect growth regulator (IGR): A substance, natural or synthetic, that controls or modifies insect growth processes.

Insecticide resistance: Genetically inherited ability to withstand doses of pesticide which would kill individuals from strains whose ancestors had not been exposed to the pesticide.

Insect resistant (plants): Tolerant of, or resistant to, insect attack (as in plants). individuals from strains whose ancestors had not been exposed to the pesticide.

Instar: The stage of an insect's life between successive molts, for example the first instar is between hatching from the egg and the first molt.

Integrated pest management (IPM): An approach to the management of pests in which all available control options, including physical, chemical, and biological controls, are evaluated and integrated into a unified program.

Integument: The outer covering of the insect body that includes the cuticle and the epidermis.

Introduction (classical biological control): The importation of a natural enemy from a foreign country or continent, usually to control a pest also of foreign origin.

Inundative release: The release of relatively large numbers of natural enemies to suppress pest populations, without the expectation that the natural enemies will colonize and spread throughout the area.

Larva, Larvae (pl.): The immature stage between the egg and pupa of insects having complete metamorphosis where the immature differs radically from the adult (e.g., caterpillars, grubs).

Leafy greens: Lettuces and other leaf vegetables.

Least toxic: Having a minimal toxic effect upon non-target organisms.

Life Cycle: The sequence of events that occurs during the lifetime of an individual organism.

Maggot: The immature form (larva) of a fly or wasp, lacking legs and a well-developed head.

Mass-reared: Produced in large numbers, as in natural enemies produced for release programs.

Mechanical control: Control of pests by physical means such as the use of screens or row covers.

Metabolism, Metabolic: Chemical changes that occur in living cells to provide energy for vital activities and to assimilate new material.

Metamorphosis: A change in body form during development of an insect.

Microbial: A microscopic organism; a germ.

Microbial insecticide: A preparation of microorganisms (e.g., viruses or bacteria) or their products used to suppress insect pest populations.

Microsporidia: Single-celled life forms, related to Protozoa.

Mite: Any of several minute invertebrates belonging to the phylum Arthropoda, class Arachnida.

Morphology: Form or structure of an organism.

Multivoltine: Having more than one brood or generation per season.

Mycelium, Mycelia (pl.): A mass of interwoven filamentous 'threads' that make up the vegetative part of a fungus.

Native (insect or plant): Of local origin, not intentionally or accidentally introduced.

Natural control: The suppression of pest populations by naturally occurring biological and environmental agents.

Natural enemies: Living organisms found in nature that kill, weaken, or reduce the reproductive potential of other organisms.

Nectar: The sugary liquid secreted by many flowers.

Nematode: An elongated, cylindrical worm parasitic in animals, insects, or plants, or free-living in soil or water.

NPV: Nuclear polyhedrosis virus.

Nymph: The immature stage, following hatching from the egg, of an insect that does not have a pupal stage.

Ocelli: Simple eyes on some adult and larval insects. Typically there are three, which form an inverted triangle dorsally, the function of which is obscure.

Order: A taxonomic subdivision that contains groups of related families or superfamilies; usually ending in -ptera in insects.

Overwinter: A period of rest or hibernation by which insects survive the winter.

Oviposition: The laying or depositing of eggs.

Ovipositor: The egg-laying apparatus of a female insect.

Palp (palpus): Sensory organ attached to insect mouthparts that is used to test the quality of food.

Parasite: An organism that lives in or on another organism (the host) during some portion of its life cycle.

Parasitoid: An animal that feeds in or on another living animal, consuming all or most of its tissues and eventually killing it.

Parthenogenesis: Development of an insect, from egg to adult, without fertilization.

Pathogen: A disease-causing organism.

Pest: An organism that interferes with human activities, property, or health, or is objectionable.

Pest management: see **Integrated pest management**.

Pesticide: A substance that is used to kill, debilitate, or repel a pest.

Pest-resistant crops: Crops that possess attributes which minimize damage by pests.

Phenology: The seasonal life history of an insect population.

Pheromone: A substance, such as a sex attractant, that is given off by one individual and causes a specific reaction in other individuals of the same species.

Phylum, Phyla (pl.): One of the major divisions of the animal kingdom.

Physical Control: Control of pests by physical means such as heat, cold, sound waves, etc.

Polyembryonic (eggs): A single egg that divides to form two or more (often hundreds) identical embryos.

Polyembryony: Having several embryos.

Population: A group of individuals of the same species within a given space and time.

Predaceous: Preying upon other organisms, predatory.

Predator: An animal that attacks and feeds on other animals, normally killing several individuals during its life cycle.

Pronotum: The upper, often shield-like, hardened body-wall plate, located just behind the head of an insect.

Protozoan: A microscopic, single-celled organism that is largely aquatic and includes many parasitic forms.

Pupa, Pupae (pl.): The nonfeeding stage between the larva and adult in insects with complete metamorphosis.

Puparium, Puparia (pl.): A case formed by the hardening of the last larval skin, in which the pupa is formed; usually of flies.

Pupate: To transform to a pupa.

Resistance (insecticide or pesticide): see **Insecticide resistance**.

Resistance (plant): see **Host plant resistance**.

Resurgence (pest): The development of large populations of pests that had previously been suppressed.

Sampling: Estimating the density of organisms (pests or natural enemies) or damage by examining a defined portion of the crop.

Scouting, Scout: see **Sampling**.

Septicemia: Blood poisoning caused by pathogenic organisms.

Specialist: A pest or natural enemy that utilizes a narrow range of species for its host or prey.

Species: A group of individuals similar in structure and capable of interbreeding and producing fertile offspring. They are different in structure from other such groups and do not interbreed with them.

Spiracles: The external openings of the insect breathing (tracheal) system, found along the abdomen.

Spore: A reproductive structure developing in certain bacteria and fungi which is strongly resistant to environmental influences but which will become active under suitable conditions.

Stage (life stage): A distinct period in the development of an organism (e.g., for some insects, egg, larval, pupal, and adult stages).

Stomach poison: An insecticide that is lethal only after it has been ingested by an insect, entering the insect body through the gut.

Systemic insecticide: An insecticide that is absorbed into plant sap and is lethal to insects feeding on or within the treated plant.

Thorax: The insect body region behind the head which bears the legs and wings.

Tolerance (host-plant resistance): The ability of a plant to withstand injury by pests.

Transformed (Bt-transformed): Transfer and expression of a gene (e.g., for Bt toxin) into another organism.

Trap crop: A small area of a crop used to divert pests from a larger area of the same or another crop. The pests, once diverted to the trap crop, may be treated with an insecticide.

Univoltine: Having only one brood or generation per season.

Vector: An organism capable of carrying and transmitting a disease-causing agent from one host to another.

Virus: Any of various submicroscopic pathogens which can only replicate inside a living cell.

Bibliography

Bauer L.S., H. Liu, D. Miller, J. Gould. 2008. Developing a classical biological control program for Agrilus planipennis (Coleoptera: Buprestidae), an invasive ash pest in *North America. Newsletter of the Michigan Entomological Society* 53: 38-39

Burks, B. D. 1954. Parasitic wasps of the *Catolaccus* group in the Americas. USDA, *Technical Bull. No. 1093*: 1-21.

Cappaert, D., and D.G. McCullough. 2009. Occurrence and seasonal abundance of *Atanycolus cappaerti* (Hymenoptera: Braconidae), a native parasitoid of emerald ash borer. *Agrilus planipennis* (Coleoptera: Buprestidae). *Gt. Lakes Entomol.* 42: 16-29

Cate, J. R. 1987. A method of rearing parasitoids of boll weevil without the host plant. Southwest. *Entomol.* 12: 211-215.

Cate, J. R., P. C. Krauter and K. E. Godfrey. 1990. Pests of cotton, pp. 17-29. En D. H. Habeck, F. D. Bennett, and J. H. Frank, Classical Biological Control in the Southern United States. *South. Coop. Ser. Bull.* 355. 197 pp.

Coleman, R. J., J. A. Morales-Ramos, E. G. King, and L. A. Wood. 1996. Suppression of the boll weevil in organic cotton by release of *Catolaccus grandis* as part of the Southern Rolling Plain boll weevil eradication program, p. 1094. In *Proceedings Beltwide Cotton Conferences 1996*, Vol. 2, National Cotton Council of America, Memphis, TN.

Coleman, R. J., S. M. Greenberg and E. G. King. 1995. Dispersal of female *Catolaccus grandis* in cotton, pp. 34-36. In *Addendum to the Proceedings Beltwide Cotton Conferences 1995*, National Cotton Council of America, Memphis, TN.

Cross, W. H. and H. C. Mitchell. 1969. Distribution and importance of *Heterolaccus grandis* as a parasite of the boll weevil. *Ann. Entomol. Soc.* Am. 62: 235-236.

Dunn, J.P., T.W. Kimmerer, G.L. Nordin. 1986. The role of host tree condition in attack of white oaks by the twolined chestnut borer, *Agrilus bilineatus* (Weber) (Coleoptera: Buprestidae). *Oecologia* 70: 596-600

Edwards, R. H., J. A. Morales-Ramos, M. G. Rojas, and E. G. King. 1998. Proposed transformation of laboratory to commercial-scale mass rearing of *Catolaccus grandis* (Hymenoptera: Pteromalidae). *Vedalia* 5: 133-139.

Elzen, G. W., M. G. Rojas, P. J. Elzen, E. G. King, and N. M. Barcenas. 1998. Toxicity of insecticides to the boll weevil parasitoid *Catolaccus grandis* (Hymenoptera: Pteromalidae). *Vedalia* 5: 123-131.

Flanders, Kathy L. and Radcliffe, Edward B. (1996) Radcliffe's IPM World Textbook: Alfalfa IPM, University of Minnesota, National IPM Network, Consortium for International Crop Protection.

Godfray, H. C. J. 1994. *Parasitoids, Behavioral and Evolutionary Ecology*. Princeton University Press. Princeton, MA.

Heinz, K. M. 1998. Host size-dependent sex allocation behavior in a parasitoid: implications for *Catolaccus grandis* (Hymenoptera: Pteromalidae) mass rearing programmes. *Bull. Entomol. Res*. 88: 37-45.

Herms D.A., D.G. McCullough, D.R. Smitley, C.S. Sadof, R.C. Williamson, P.L. Nixon, 2009. Insecticide options for protecting ash trees from emerald ash borer. *North Central IPM Center Bulletin*. 12 pp.

Hogg, David B. (1994) Know Your Friends: Bathyplectes, Parasites of the Alfalfa Weevil, *Midwest Biological Control News Online*. Vol. I, No.1.

Johnson, W. L., W. H. Cross, W. L. McGovern and H. C. Mitchell. 1973. Biology of *Heterolaccus grandis* in a laboratory culture and its potential as an introduced parasite of the boll weevil in The United States. *Environ. Entomol*. 2: 112-118.

King, E. G., R. J. Coleman, L. Woods, L. Wendel, S. Greenberg, and A. W. Scott. 1995. Suppression of the boll weevil in commercial cotton by augmentative releases of a wasp parasite, *Catolaccus grandis*, pp. 26-30. In Addendum to the *Proceedings Beltwide Cotton Conferences 1995*, National Cotton Council of America, Memphis, TN.

Marsh, P.M., J.S. Strazanac, S.Y. Laurusonis. 2009. Description of a new species of *Atanycolus* (Hymenoptera: Braconidae) from Michigan reared from the emerald ash borer. *Gt. Lakes Entomol.* 42: 8-15

Morales-Ramos, J. A. and E. G. King. 1991. Evaluation of *Catolaccus grandis* (Burks) as a biological control agent against the cotton boll weevil, p. 724. In *Proceedings Beltwide Cotton Conferences 1991*, Vol. 2, National Cotton Council of America, Memphis, TN.

Morales-Ramos, J. A. and J. R. Cate. 1992a. Laboratory determination of age-dependent fecundity, development, and rate of increase of *Catolaccus grandis* (Burks) (Hymenoptera: Pteromalidae). *Ann. Entomol. Soc. Am.* 85: 469-476.

Morales-Ramos, J. A. and J. R. Cate. 1992b. Functional response of *Catolaccus grandis* (Burks) (Hymenoptera: Pteromalidae) in field cages. *Biol. Cont.* 2: 193-202.

Morales-Ramos, J. A. and J. R. Cate. 1993. Temperature-dependent developmental rates of *Catolaccus grandis* (Burks) (Hymenoptera: Pteromalidae). *Environ. Entomol.* 22: 226-233.

Morales-Ramos, J. A., K. R. Summy and E. G. King. 1995b. Estimating parasitism by *Catolaccus grandis* (Hymenoptera: Pteromalidae) after inundative releases against the boll weevil (Coleoptera: Curculionidae). *Environ. Entomol.* 24: 1718-1725.

Morales-Ramos, J. A., K. R. Summy, J. L. Roberson, J. R. Cate, and E. G. King. 1992. Feasibility of mass rearing of *Catolaccus grandis,* a parasitoid of the boll weevil, pp. 723-726. In *Proceedings Beltwide Cotton Conferences 1992,* Vol. 2, National Cotton Council of America, Memphis, TN.

Morales-Ramos, J. A., M. G. Rojas and E. G. King. 1995a. Venom of *Catolaccus grandis* (Hymenoptera: Pteromalidae) and its role in parasitoid development and host regulation. *Ann. Entomol. Soc. Am.* 88: 800-808.

Morales-Ramos, J. A., M. G. Rojas, and E. G. King. 1996. Significance of adult nutrition and oviposition experience on longevity and attainment of full fecundity of *Catolaccus grandis* (Hymenoptera: Pteromalidae). *Ann. Entomol. Soc. Am.* 89: 555-563.

Morales-Ramos, J. A., M. G. Rojas, J. Roberson, R. G. Jones, E. G. King, K. R. Summy and J. R. Brazzel. 1994. Suppression of the boll weevil first generation by augmentative releases of *Catolaccus grandis* in Aliceville, Alabama, pp. 958-964. In *Proceedings Beltwide Cotton Conferences* 1994, Vol.2, National Cotton Council of America, Memphis, TN.

Morales-Ramos, J. A., M. G. Rojas, R. J. Coleman and E. G. King. 1998. Potential use of in vitro-reared *Catolaccus grandis* (Hymenoptera: Pteromalidae) for biological control of the boll weevil (Coleoptera: Curculionidae). J. Econ. Entomol. 91: 101-109.

Paetel, C. 1997. Investigations on potential biocontrol agents of Dalmatian and yellow toadflax, *Linaria dalmatica* and *L. vulgaris*. *Intl. Inst. of Biol. Control European Station Annual Report*. 22 pp.

Quicke, D. L. J. 1997. *Parasitic Wasps*. Chapman and Hall, New York.

Rojas, M. G., J. A. Morales-Ramos and E. G. King. 1996. *In vitro* rearing of the boll weevil (Coleoptera: Curculionidae) ectoparasitoid *Catolaccus grandis* (Hymenoptera: Pteromalidae) on meridic diets. *J. Econ. Entomol.* 89: 1095-1104.

Rojas, M. G., J. A. Morales-Ramos, and E. G. King. 1999. Response of *Catolaccus grandis* (Hymenoptera: Pteromalidae) to its Natural Host After Ten Generations of Rearing on a Factitious Host (*Callosobrucus maculatus*) (Coleoptera: Bruchidae). *Environ. Entomol*. 28: 137-141.

Rojas, M. G., J. A. Morales-Ramos, E. G. King, G. Saldaña and S. M. Greenberg. 1998. Use of a factitious host and supplemented adult diet to rear and induce oogenesis in *Catolaccus grandis* (Hymenoptera: Pteromalidae). *Environ. Entomol*. 27: 499-507.

Smith, J.M. 1959. Notes on insects, especially *Gymnaetron* spp. (Coleoptera: Curculionidae), associated with toadflax, *Linaria vulgaris* Mill. (Scrophulariaceae), in North America. *Can. Entomol*. 91: 116-121.

Summy, K. R., J. A. Morales-Ramos and E. G. King. 1995a. Suppression of boll weevil infestations on South Texas cotton by augmentative releases of the exotic parasite *Catolaccus grandis* (Hymenoptera: Pteromalidae). *Biological Control* 5: 523-529.

Summy, K. R., J. A. Morales-Ramos, E. G. King, R. J. Coleman and A. W. Scott. 1994b. Impact of mechanical cultivation on the searching efficiency of *Catolaccus grandis*, an exotic parasite of boll weevil. Southwest. *Entomol*. 19: 379-384.

Summy, K. R., J. A. Morales-Ramos, E. G. King, S. Greenberg, and R. J. Coleman. 1994 a. Integration of boll weevil parasite augmentation into the short-season production system of the Lower Rio Grande Valley, pp. 953-957. *In Proc. Beltwide Cotton Conferences* 1994 Vol. 2. National Cotton Council of America, Memphis, TN.

Summy, K. R., J. A. Morales-Ramos, E. G. King, S. M. Greenberg, M. A. Rankin, L. G. Hansen and C. Moomaw. 1995b. Rationale and potential for augmentative biological control of boll weevil on follow-season cotton in southern Texas. Southwest. *Entomol*. 20: 483-491.

Summy, K. R., S. M. Greenberg, J. A. Morales-Ramos, and E. G. King. 1997. Suppresion of boll weevil infestations (Coleoptera: Curculionidae) occurring on follow-season cotton in Southern Texas by augmentative releases of *Catolaccus grandis* (Hymenoptera: Pteromalidae). *Biological Control* 9: 209-215.

Vargas-Camplis, J., R. J. Coleman, J. Gonzalez and L. Rodriguez Del B. 1997. Life table analysis of cotton boll weevil in the tropics of Tamaulipas Mexico after *Catolaccus grandis* releases, pp. 1194-1197. *In Proceedings Belwide Cotton Conferences*, 1997 Vol. 2. National Cotton Council of America, Memphis, TN.

Waage, J. and D. Greathead. 1986. *Insect Parasitoids*. Academic Press, New York.